CALCULUS

Using

Maple

CALCULUS

Using

Maple

EDWARDS & PENNEY

To Accompany

CALCULUS WITH ANALYTIC GEOMETRY

Fourth Edition

Prentice Hall
Englewood Cliffs, New Jersey 07632

Editorial/production supervision: bookworks
Acquisitions editor: George Lobell
Development editor: Karen Karlin
Managing editor: Jeanne Hoeting
Buyer: Gertrude Pisciotti

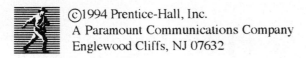

Printed in the United States of America

10 9 8 7 6 5 4 3 2 1

ISBN 0-13-458282-9

Prentice-Hall International (UK) Limited, *London*
Prentice-Hall of Australia Pty. Limited, *Sydney*
Prentice-Hall Canada Inc., *Toronto*
Prentice-Hall Hispanoamericana, S.A., *Mexico*
Prentice-Hall of India Private Limited, *New Delhi*
Prentice-Hall of Japan, Inc., *Tokyo*
Simon & Schuster Asia Pte. Ltd., *Singapore*
Editora Prentice-Hall do Brasil, Ltda., *Rio de Janeiro*

Contents

Techniques of Integration

Polar Coordinates and Conic Sections

Infinite Series

Parametric Curves and Vectors in the Plane

Vectors, Curves, and Surfaces in Space

Partial Differentiation

Multiple Integrals

Vector Analysis

Preface

There is wide interest in the idea of calculus as a laboratory course. This is a laboratory manual written to accompany Edwards & Penney, *Calculus with Analytic Geometry*, Prentice Hall, 4th edition (1994). It consists of versions of the textbook's 48 projects that have been expanded where necessary to take advantage of specific computational technologies in the teaching of calculus.

Although each project references the appropriate section of Edwards & Penney, the discussion within each project is independent of any particular text. The manual therefore can be used in conjunction with any mainstream calculus textbook that spans the usual range of topics from functions and graphs to multiple integrals.

Each project is designed to provide the basis for an outside class assignment that will occupy students for a period of several days (or perhaps a week or two). The various projects may be used in a variety of ways, depending on the specific technology that is available and on the local arrangements for its use. We ourselves have experimented with several formats ranging from brief daily homework assignments to intensive projects leading to careful written reports.

At the present time, *Derive*, *Maple*, *Mathematica*, *MATLAB*, *X(PLORE)*, and Hewlett-Packard and Texas Instruments graphics calculator editions of this manual are available. The *Derive*, *Maple*, and *Mathematica* versions are accompanied by appropriate command-specific computer diskettes. The *Maple* diskette contains a *Maple* worksheet for each of the manual's 48 projects. These diskettes will relieve students of much of the burden of typing by providing "templates" for the principal commands used in each project. In many cases, the diskettes also contain additional discussion and examples.

As both computational technology and its uses in the teaching of calculus mature, further editions and technology-specific versions of this manual will be prepared. We will appreciate faculty suggestions as to appropriate revisions and new projects for inclusion, as well as student comments on how our project discussions can be improved for lab use or independent study. These suggestions can be forwarded to us at the following address:

Henry Edwards & David Penney
Department of Mathematics
University of Georgia
Athens, GA 30602-7403

hedwards@math.uga.edu
dpenney@math.uga.edu

CALCULUS

Using

Maple

Chapter 1

Functions and Graphs

Project 1
Solution of Equations by the Method of Tabulation

Reference: Section 1.1 of Edwards & Penney

In this project you will solve equations using *Maple*'s ability to construct tables of values of functions. For instance, suppose you want to solve the equation $f(x) = 0$. The *Maple* command

```
for x from a by h to b do print f(x) od;
```

generates a list of values of the function $f(x)$ on the interval $[a, b]$, proceeding from $x = a$ to $x = b$ by steps of length h. You can then scan this list of values to see where sign changes bracketing a solution of the equation occur. The more elaborate form

```
for x from a by h to b do
      print (x,`          `,f(x))        #  note backquotes
      od;
```

of this command generates a nice two-column table with values of x in the first column and values of $f(x)$ in the second column.

Example
The quadratic equation

$$x^2 - 2x - 5 = 0 \tag{1}$$

has two solutions that could be found by using the quadratic equation of high school algebra. Instead, we tabulate! If we first define the function on the left-hand side by entering the *Maple* command

```
f := x -> x^2 - 2*x - 5;
```

then the command

```
for x from -5 by 1 to 5 do
      print (x,`        `,f(x))
      od;
```

generates the table

-5	30
-4	19
-3	10

1

```
-2        3
-1       -2
 0       -5
 1       -6
 2       -5
 3       -2
 4        3
 5       10
```

with values of x in the first column and corresponding values of $f(x)$ in the second column. Scanning the signs in the second column, we see that $f(x)$ changes sign in the x-intervals [-2, -1] and [3, 4]. Hence the two roots of Eq. (1) lie in these two intervals. One root is positive and the other is negative. To approximate the positive solution, for instance, we tabulate $f(x)$ in the interval [3, 4].

```
for x from 3 by 0.1 to 4 do
      print (x,`       `,f(x))
      od;


3             -2
3.1           -1.59
3.2           -1.16
3.3           -0.71
3.4           -0.24
3.5           0.25
3.6           0.76
3.7           1.29
3.8           1.84
3.9           2.41
4.            3.
```

Now we see that our solution lies in [3.4, 3.5], and another tabulation gives

```
for x from 3.4 by 0.01 to 3.5 do
      print (x,`        `  ,f(x))
      od;


3.4           -0.24
3.41          -0.1919
3.42          -0.1436
3.43          -0.0951
3.44          -0.0464
3.45          0.0025
3.46          0.0516
3.47          0.1009
3.48          0.1504
3.49          0.2001
3.5           0.25
```

At this point it is clear (why?) that the positive solution of Eq. (1) is $x \approx 3.45$, accurate to 2 decimal places. An additional decimal place of accuracy can be obtained with each additional "subtabulation." You should practice this method by finding similarly the negative solution of Eq. (1). Then compare the results with those obtained by use of the quadratic formula $x = \left(-b \pm \sqrt{b^2 - 4ac}\right) / (2a)$.

As a warm-up for the projects below, you can apply the method of repeated tabulation to approximate (accurate to three or four decimal places) the two (real) roots of each of the following quadratic equations.

$$x^2 - x - 3 = 0$$
$$x^2 + 2x - 4 = 0$$
$$2x^2 + 3x - 10 = 0$$
$$4x^2 - 7x - 47 = 0$$

Project 1A

Suppose that you need to find the (positive) square root of 2 accurate to three decimal places, but your calculator has no square root key, only keys for the four arithmetic operations $+, -, \times, \div$. Could you nevertheless approximate $\sqrt{2}$ accurately using such a simple calculator? Here you will explore this question, for convenience using *Maple* rather than a hand-held calculator.

You are looking simply for a number x such that $x^2 = 2$; that is, such that $x^2 - 2 = 0$. Thus $x = \sqrt{2}$ is a solution of the equation

$$f(x) = x^2 - 2 = 0.$$

And even with a simple four-function calculator, you could easily calculate any desired value of $f(x)$: Merely multiply x by x, then subtract 2. Consequently you could readily tabulate values of the function $f(x) = x^2 - 2$ with very simple calculations.

So apply the method of repeated tabulation to approximate the number $\sqrt{2}$ accurate to three decimal places. Similarly, you can apply the method of repeated tabulation to find three-place approximations to such roots as

- $\sqrt{17}$ as a solution of $x^2 = 17$,

- $\sqrt[3]{25}$ as a solution of $x^3 = 25$,

- $\sqrt[5]{100}$ as a solution of $x^5 = 100$.

Project 1B

Figure 1 shows a 50-ft by 100-ft rectangular plot that you plan to enclose with a sidewalk of width x costing 25¢ per square foot. If you have $250 to pay for the sidewalk, determine the value of x accurate to 0.01 ft. (That is, determine the largest value of x so that the cost does not exceed $250.) First express the area of the sidewalk as the difference of the areas of the outer and inner rectangles in Fig. 1. Then show that

$$(2x + 100)(2x + 50) - 5000 = 1000.$$

Finally, approximate x by repeated tabulation.

3

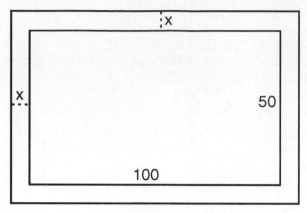

Figure 1

For some more interesting possibilities, you might replace the rectangular plot with

- An L-shaped plot,
- A plot shaped like an isosceles right triangle, or an equilateral triangle, or a 3-4-5 right triangle,
- A plot shaped like a Norman window (a rectangle beneath a semicircle),
- A regular hexagon.

For comparable results, let the plot in each case have a perimeter of approximately 300 ft.

Project 2
Solution of Equations by the Method of Successive Zooms

Reference: Section 1.3 of Edwards & Penney

The projects here involve the use of *Maple* to magnify the graph $y = f(x)$ so as to "zoom in" on its intersections with the *x*-axis and thereby locate (at least approximately) the solutions of the equation $f(x) = 0$. This is the *method of repeated magnification* or "zooming" for the graphical solution of equations. The command

```
plot( f(x), x = a..b, y = c..d );
```

plots the portion of the graph $y = f(x)$ that lies within the "window" $a \le x \le b$, $c \le y \le d$ in the *xy*-plane.

Example
The quadratic equation

$$x^2 - 2x - 5 = 0 \tag{1}$$

4

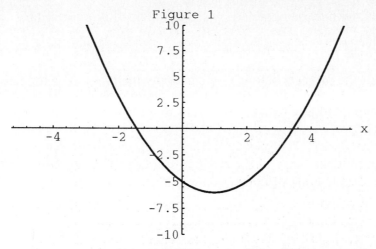

Figure 1

has two solutions that could be found by using the quadratic equation of high school algebra. Instead, we zoom! If we first define the function on the left-hand side by entering the *Maple* command

```
f := x -> x^2 - 2*x - 5;
```

then the command

```
plot( f(x), x = -5..5, y = -10..10 );
```

produces the graph shown in Fig. 1.

We see solutions of Eq. (1) in the intervals $[-2, -1]$ and $[3, 4]$. To investigate the positive root on the right, for instance, we execute the command

```
plot( f(x), x = 3..4, y = -1..1 );
```

and get the graph shown in Fig. 2. Another zoom produces Fig. 3, which makes it clear that the positive solution of Eq. (1) is $x \approx 3.45$, accurate to two decimal places.

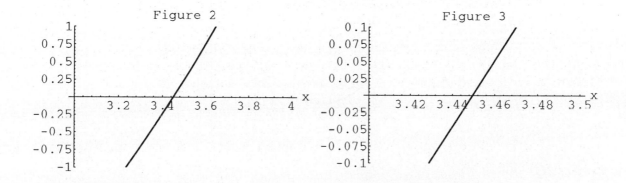

Figure 2   Figure 3

An additional decimal place of accuracy can be obtained with each additional zoom. You should practice this method by finding similarly the negative solution of Eq. (1).

5

Then compare the results with those obtained by use of the quadratic formula,
$$x = \left(-b \pm \sqrt{b^2 - 4ac} \right) / (2a).$$

As a warm-up for the projects below, you can apply the method of repeated magnification to approximate (accurate to three or four decimal places) the two (real) roots of each of additional quadratic equations such as those listed in Project 1A.

Project 2A

Project 1A discusses the number $\sqrt{2}$ as the positive solution of the equation $x^2 - 2 = 0$, which is the same as the intersection of the parabola $y = x^2 - 2$ and the positive x-axis. Magnify successively the graph $y = x^2 - 2$ so as to approximate $\sqrt{2}$ accurate to four decimal places.

Similarly, you might apply the method of successive zooms to find four-place approximations to such roots as $\sqrt{17}, \sqrt[3]{25},$ and $\sqrt[5]{100}$.

Project 2B

Project 1B deals with a sidewalk of width x feet that borders a 50-ft by 100-ft rectangular plot. Conclude from the discussion there that, if the total area of the sidewalk alone is to be 1000 ft^2, then x must be the x-coordinate of a point of intersection of the line $y = 1000$ and the parabola $y = (2x + 100)(2x + 50) - 5000$. The command

```
plot( {1000, (2*x + 100)*(2*x + 50) - 5000},
            x = -90..20, y = -7000..3000 );
```

plots the line and the parabola simultaneously (Fig. 4). Apply the method of successive zooms to find (accurate to four decimal places) each of the two evident intersection points. Are there actually two possible values of x?

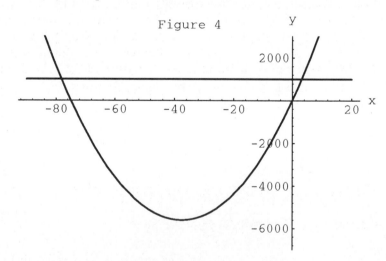

Figure 4

Project 2C

Figure 5 shows a 12-ft ladder leaning across a 5-ft fence and touching a high wall located 3 ft behind the fence. We want to find the distance x from the base of this ladder to the bottom of the fence. Write up carefully the following discussion: Application of the Pythagorean theorem to the large right triangle in Fig. 5 yields

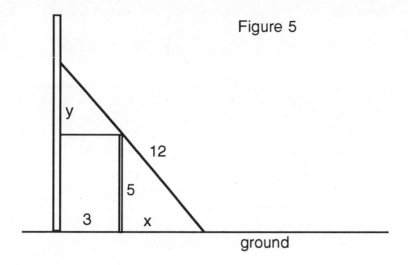

Figure 5

$$(x + 3)^2 + (y + 5)^2 = 144, \qquad (2)$$

the equation of a circle with center $(-3, -5)$ and radius 12. Then note that the two small triangles in the figure are similar. Hence $y/3 = 5/x$, so

$$y = \frac{15}{x} \qquad (3)$$

Show that the circle in Eq. (2) and the graph of Eq. (3) intersect in four points -- two in the first quadrant and two in the third quadrant. The two first-quadrant intersection points yield *two* physically possible positions of the ladder. Apply the method of successive zooms to find them, with x and y accurate to four decimal places. Sketch, roughly to scale, these two physical possibilities. Why do the two third-quadrant intersection points not yield two additional physically possible positions of the ladder?

Project 3
More Solution of Equations by Zooming

Reference: Section 1.4 of Edwards & Penney

Each of the problems below calls for the graphical solution of one or more equations by the method of successive magnifications (or zooms) that was introduced in Project 2. Determine each desired solution accurate to three decimal places. Figures 1 and 2 illustrate, respectively, the three solutions of a typical cubic equation and the four solutions of a typical quartic (fourth degree) equation. In each case finding all real solutions of the equation $f(x) = 0$ requires finding all x-intercepts of the graph $y = f(x)$.

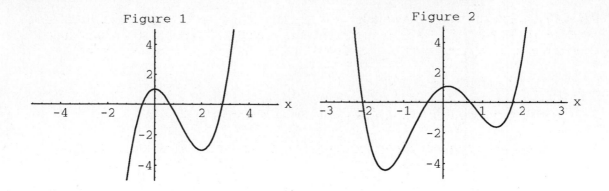

Figure 1   Figure 2

1.   Find all real solutions of the cubic equations

   (a) $x^3 - 3x^2 + 1 = 0$;   (Fig. 1)

   (b) $x^3 - 3x^2 - 2 = 0$.

2.   Find all real solutions of the quartic equation

   $x^4 - 4x^2 + x + 1 = 0$.   (Fig. 2)

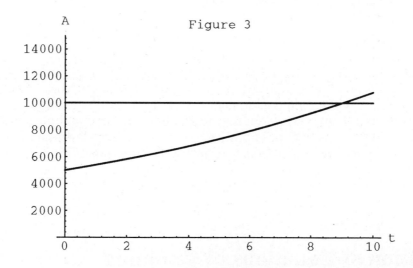

Figure 3

3.   (a) Suppose that you invest $5000 in an account that pays interest compounded continuously, with an annual interest rate of 7.696%, so that the amount on deposit at time t (in years) is given by

   $A(t) = 5000 \cdot 1.08^t$.

Beginning with Fig. 3, generated with the *Maple* command

```
plot( {10000, 5000*(1.08)^t},
            t = 0..10, A = 0..15000 );
```

8

determine graphically how long it takes -- to the nearest day -- for your initial investment of $5000 to double. (b) If the interest rate were, instead, 9.531% but still compounded continuously, then the amount you would have on deposit after t years would be

$$A(t) = 5000 \cdot (1.10)^t.$$

Find graphically how long it would take for your investment to *triple*.

4.   Suppose that a population is described by an exponential function of the form $P(t) = P_0 2^t$.

(a)   If this population doubles in six months, how long does it take to triple?
(b)   If this population triples in six months, how long does it take to double?

5.   Find all real solutions of the equations

         (a) $x = \cos x$ (Fig. 4);

         (b) $x^2 = \cos x$;

         (c) $1 - x = 3 \cos x$ (Fig. 5).

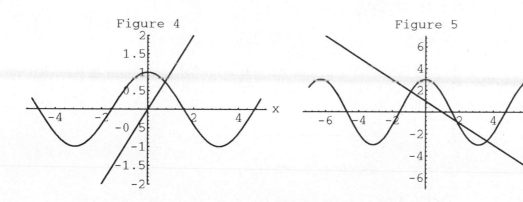

6.   Find all *positive* solutions of the equations

         (a) $2^x = 3 \cos x$;

         (b) $2^x = 3 \cos 4x$.

Chapter 2

Prelude to Calculus

Project 4
Graphical Approximation
of Slopes of Curves

Reference: Section 2.1 of Edwards & Penney

This project is based on the the fact that a curve and its tangent line are virtually indistinguishable at a sufficiently high level of magnification. We can therefore approximate the slope of a curve at a point P by magnifying the curve near P, selecting a pair of points lying on either side of P, and then calculating the slope of the line through these selected points. Here we describe how to use the facilities of *Maple* to do this efficiently.

Example
Suppose we want to approximate the slope of the curve $y = x^2$ (that is, of its tangent line) at the point $P(1, 1)$. Defining

```
f := x -> x^2;
```

our idea is to plot the line through the points $A(1 - h, f(1 + h))$ and $B(1 + h, f(1 + h))$ on either side of P. Figure 1 shows the parabola and this straight line (with $h = 0.5$) plotted with the commands

```
h := 0.5;
x0 := 1;
m := (f(x0+h) - f(x0-h))/(2*h);
a := x0 - h;
plot( {f(x), f(a) + m*(x-a)}, x = x0-2*h..x0+2*h );
```

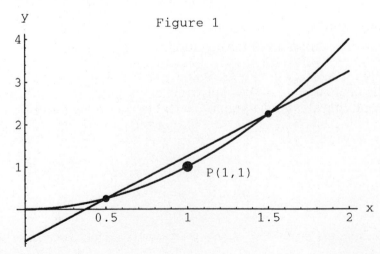

Figure 1

10

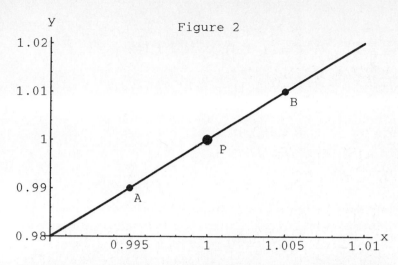

Figure 2

Figure 2 shows the same plot, except now with $h = 0.005$ instead of 0.5. Note that even at this relatively low level of magnification, the portion of the parabola shown is visually indistinguishable from the straight line through the points A and B on either side of P.

We can use the slope m of the straight line as an approximation to the slope of the tangent line to the the parabola at the point P. This approximating slope is readily calculated by the self-contained commands

```
h := 0.005;
x0 := 1;
deltaY := f(x0 + h) - f(x0 - h);
deltaX := 2*h;
slope := deltaY/deltax;
```

$$\text{slope} := 2.000000000$$

As in this example, the "two-sided difference quotient"

$$\frac{f(x+h)-f(x-h)}{2h}$$

is ordinarily an excellent approximation to the numerical value of the actual derivative $f'(x)$.

Each problem below lists a curve $y = f(x)$ and a point $P(a, f(a))$ at which the slope $f'(a)$ of the tangent line is to be approximated graphically. Zoom in on P (magnifying several times) until the graph $y = f(x)$ looks (in the resulting viewing window) precisely like a straight line. At each zoom, record the coordinates (x_1, y_1) and (x_2, y_2) of two points located on each side of P (as in Fig. 2). Then let

$$m_k = \frac{\Delta y}{\Delta x} = \frac{y_2 - y_1}{x_2 - x_1}$$

denote the approximate slope that results at the kth zoom. Is it clear what limiting value the approximate slopes m_1, m_2, m_3, ... are approaching? (In each case $f'(x)$ should be an integer or an integral multiple of 1/2, 1/4, or 1/8.)

1. $f(x) = x^2$, $P = P(-2, 4)$; $f'(-2) = ?$

2. $f(x) = \sqrt{x}$, $P = P(1, 1)$; $f'(1) = ?$

3. $f(x) = \dfrac{1}{x}$, $P = P(2, 1/2)$; $f'(2) = ?$

4. $f(x) = \dfrac{12}{x^2}$, $P = P(-4, 3/4)$; $f'(-4) = ?$

5. $f(x) = \sqrt{x^2 - 9}$, $P = P(5, 4)$; $f'(5) = ?$

Project 5
Numerical Investigation of Limits

Reference: Section 2.2 of Edwards & Penney

The object of this project is the systematic use of a computer for the numerical investigation of limits. Suppose that we want to investigate the value (if any) of the limit

$$\lim_{x \to a} f(x)$$

of a given function f at $x = a$. We shall begin with a fixed increment h, then calculate (as efficiently as possible) the values of f at the points

$$a + \frac{h}{5}, a + \frac{h}{5^2}, \ldots, a + \frac{h}{5^n}, \ldots$$

that approach the number a as n increases.

To investigate a limit as $x \to 0$, for instance, we might take $a = 0$ and $h = 1$, then calculate the numerical value $f(x)$ at the points

$$0.2, \ 0.04, \ 0.008, \ \ldots, \ (0.2)^n, \ \ldots$$

that approach zero as n increases.

Example
With the function

$$f(x) = \frac{\sqrt{x + 25} - 5}{x},$$

the *Maple* commands

12

```
a := 0;    h := 1;
x := n -> a + h/5^n;
f := x -> ((x+25)^0.5 - 5)/x;
for n from 1 by 1 to 8 do
     print( evalf(x(n)),`        `,evalf(f(x(n))) );
     od;
```

produce the table

0.2	0.0998008
0.04	0.0999600
0.008	0.0999920
0.0016	0.0999984
0.00032	0.0999997
0.000064	0.0999999
0.0000128	0.1000000
0.00000256	0.1000000

with values of x approaching 0 in the first column and values of $f(x)$ in the second column. These numerical results suggest strongly that

$$\lim_{x \to 0} \frac{\sqrt{x+25}-5}{x} = \frac{1}{10}.$$

Investigate similarly the numerical values (and existence) of the limits given in Problems 1 through 10. You should use several different values of h (both positive and negative) in each problem.

1. $\displaystyle\lim_{x \to 0} \frac{(1+x)^2 - 1}{x}$

2. $\displaystyle\lim_{x \to 1} \frac{x^4 - 1}{x - 1}$

3. $\displaystyle\lim_{x \to 0} \frac{\sqrt{x+9} - 3}{x}$

4. $\displaystyle\lim_{x \to 4} \frac{x^{3/2} - 8}{x - 4}$

5. $\displaystyle\lim_{x \to 0} \left(\frac{1}{x+5} - \frac{1}{5} \right)$

6. $\displaystyle\lim_{x \to 8} \frac{x^{2/3} - 4}{x - 8}$

7. $\displaystyle\lim_{x \to 0} \frac{\sin x}{x}$

13

8. $\lim\limits_{x\to 0} \dfrac{2^x - 1}{x}$

9. $\lim\limits_{x\to 0} \dfrac{10^x - 1}{x}$

10. $\lim\limits_{x\to 0} \left(1 + \dfrac{1}{x}\right)^x$

Project 6
Applications of Cubic and Quartic Equations

Reference: Section 2.4 of Edwards & Penney

Both of the problems below involve a geometric problem that leads to a higher-degree equation that can be solved by the method of successive magnification (as in Projects 2 and 3). We also illustrate the use of symbolic algebra commands in *Maple* to eliminate an unwanted variable in a pair of equations.

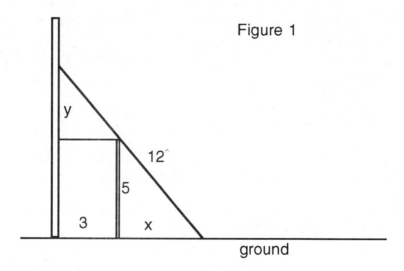

Figure 1

Project 6A
Figure 1 shows the leaning ladder of Project 2C. There you were asked to show that the indicated lengths x and y satisfy the equations

$$(x + 3)^2 + (y + 5)^2 = 144,$$

$$xy = 15.$$

      The usual strategy for solving two equations in two unknowns is to start by eliminating one of the unknowns to obtain a single equation in a single unknown. Using *Maple* we can first name our two equations, as with the commands

```
equation1 :=            (x+3)^2 + (y+5)^2 = 144;
equation2 :=                    x*y = 15;
```

Then we solve the second equation for y,

```
solve(equation2, y);
```

and we substitute the result $y = 15/x$ into the first equation and simplify by means of the successive *Maple* commands

```
equation :=     subs(y = 15/x, equation1);
equation :=     simplify(equation);
equation :=     equation*(x^2);
equation :=     equation - (144*x^2 = 144*x^2);
```

which you should execute to follow the process of simplification for yourself. The final result is the single quartic equation

$$f(x) = x^4 + 6x^3 - 110x^2 + 150x + 225 = 0 \qquad (1)$$

in the single unknown x.

Figure 2 shows the graph of the function f. First apply the intermediate value property of continuous functions to *prove* that Eq. (1) has four real solutions, and locate them in intervals whose endpoints are consecutive integers. Then use repeated tabulation or successive zooms to approximate each of these roots accurate to at least three places. What are the physical possibilities for the distance x from the bottom of the fence to the foot of the ladder?

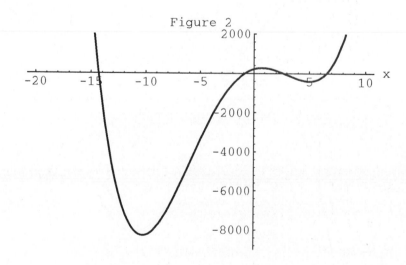
Figure 2

15

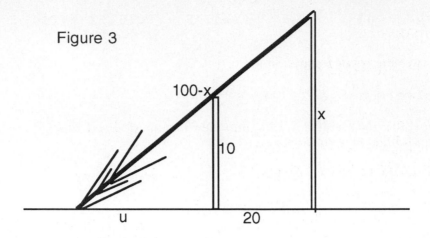

Figure 3

Project 6B

A 100-ft tree stands 20 ft from a 10-ft fence. Then (as indicated in Fig. 3) the tree is "broken" at a height of x feet. The tree falls so that its trunk barely touches the top of the fence when the tip of the tree strikes the ground on the other side of the fence. Use similar triangles and the Pythagorean theorem to show that x satisfies the cubic equation

$$f(x) = x^3 - 68x^2 + 1100x - 5000 = 0. \tag{2}$$

Figure 4 shows the graph of the function f. Apply the intermediate value property of continuous functions to *prove* that Eq. (2) has three real solutions, and locate them in intervals whose endpoints are consecutive integers. Then use repeated tabulation or successive zooms to approximate each of the roots accurate to at least three decimal places. What are the physical possibilities for the height x?

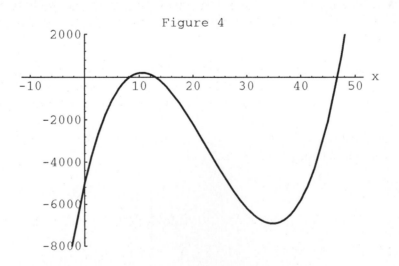

Figure 4

16

Chapter 3

The Derivative

Project 7
Graphical Investigation of Population Growth

Reference: Section 3.1 of Edwards & Penney

This project involves a graphical analysis of the population growth of a certain small city during the decade of the 1990s.

For your own small city, choose a positive integer k with $k \leq 9$ (perhaps the last nonzero digit of your student I.D. number). Then suppose that the population P of the city t years after 1990 is given (in thousands) by

$$P(t) = 10 + t - (0.1)t^2 + (0.001)(k + 5)t^3.$$

Then investigate the following questions.

1.   Does the graph of $P(t)$ indicate that the population is increasing throughout the 1990s? Explain your answer.

2.   Does the graph of the derivative $P'(t)$ confirm that $P(t)$ is increasing throughout the 1990s? What property of this graph is pertinent to the question?

3.   What points on the graph of $P'(t)$ correspond to the time (or times) at which the instantaneous rate of change of P is equal to its average rate of change between the years 1990 and 2000? Apply the method of successive magnification to find each such time (accurate to two decimal places).

4.   What points on the graph of the derivative $P'(t)$ correspond to the time (or times) at which the population $P(t)$ is increasing the slowest? The fastest? Apply the method of successive magnification to find each such time (accurate to two decimal places).

You may find it useful to plot the function $P(t)$ and its derivative $P'(t)$ simultaneously by means of the *Maple* commands

```
k := 5;

P := t -> 10 + t - 0.1*t^2 + 0.001*(k + 5)*t^3;

plot( {P(t), diff(P(t),t)}, t = 0..10 );
```

Note the *Maple* notation `diff(P(t),t)` for the derivative $P'(t)$ of the function $P(t)$ with respect to t.

Project 8
Extrema by Zooming in on Zeros of Derivatives

Reference: Section 3.5 of Edwards & Penney

Figure 1 shows the graphs of the function

$$f(x) = 4x^4 - 11x^2 - 5x - 3$$

and its derivative

$$f'(x) = 16x^3 - 22x - 5$$

on the interval $[-2, 2]$. The maximum value of $f(x)$ on $[-2, 2]$ is $f(-2) = 27$ at the left endpoint. The lowest point of the curve $y = f(x)$ corresponds to the positive zero of the derivative $dy/dx = f'(x)$.

If we attempt to zoom in on the lowest point -- without changing the "range factors" or "aspect ratio" of the viewing windows -- we get a picture as in Fig. 2, where it is difficult to locate the lowest point precisely. The reason is this: After sufficient magnification, the graph is indistinguishable from its tangent line, which is horizontal at a local maximum or minimum point.

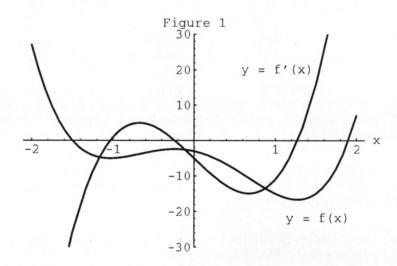

Figure 1

Consequently it is much better to zoom in on the corresponding zero of the derivative $f'(x)$. We can then locate with much greater precision the indicated critical point (Fig. 3). Here it is clear that the minimum value attained by $f(x)$ on $[-2, 2]$ is approximately $f(1.273) \approx -16.686$.

18

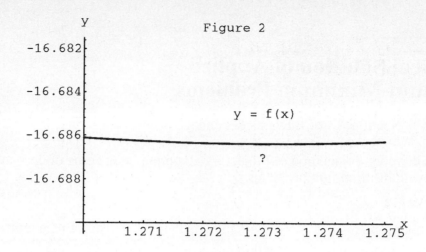

Figure 2

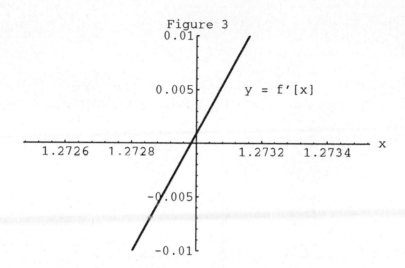

Figure 3

In Problems 1 through 8, find the maximum and minimum values of the given function on the indicated closed interval by zooming in on the zeros of the derivative.

1. $f(x) = x^3 + 3x^2 - 7x + 10;\ [-2, 2]$

2. $f(x) = x^3 + 3x^2 - 7x + 10;\ [-4, 2]$

3. $f(x) = x^4 - 3x^3 + 7x - 5;\ [-3, 3]$

4. $f(x) = x^4 - 5x^3 + 17x - 5;\ [-3, 3]$

5. $f(x) = x^4 - 5x^3 + 17x - 5;\ [0, 2]$

6. $f(x) = x^5 - 5x^4 - 15x^3 + 17x^2 + 23x;\ [-1, 1]$

7. $f(x) = x^5 - 5x^4 - 15x^3 + 17x^2 + 23x;\ [-3, 3]$

8. $f(x) = x^5 - 5x^4 - 15x^3 + 17x^2 + 23x;\ [0, 10]$

Project 9
Graphical Solution of Applied Maximum-Minimum Problems

Reference: Section 3.6 of Edwards & Penney

Here you will apply the method of Project 8 -- zooming in on zeros of derivatives -- to applied maximum-minimum problems.

Project 9A
The following problems deal with alternative methods of constructing a tent.

1.   Figure 1 shows a 20-by-20-ft square of canvas tent material. Girl Scout Troop A must cut pieces from its four corners as indicated, so that the four remaining triangular flaps can be turned up to form a tent in the shape of a pyramid with a square base. How should this be done to maximize the volume of the tent?

Let A denote the area of the base of the tent and h its height. With x as indicated in the figure, show that the volume $V = \frac{1}{3}Ah$ of the tent is given by

$$V(x) = \frac{4}{3}x^2\sqrt{100 - 20x}, \qquad 0 \le x \le 5.$$

Maximize V by graphing $V(x)$ and $V'(x)$ and zooming in on the zero of $V'(x)$.

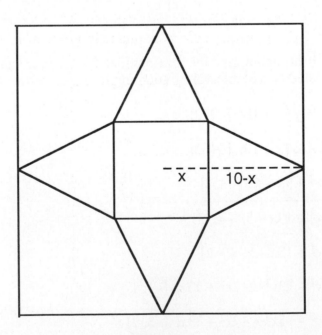

Figure 1

20

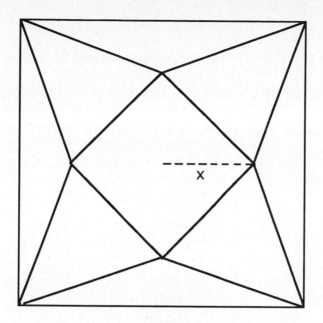

Figure 2

2.   Girl Scout Troop B must make a tent in the shape of a pyramid with a square base from a similar 20-by-20-ft square of canvas, but in the manner indicated in Fig. 2. With x as indicated in the figure, show that the volume of the tent is given by

$$V(x) = \frac{2}{3}x^2\sqrt{200 - 20x}, \quad 0 \le x \le 10.$$

Maximize V graphically as in Problem 1.

3.   Solve Problems 1 and 2 analytically to verify that the maximal volume in Problem 2 is exactly $2\sqrt{2}$ times the maximal volume in Problem 1. It pays to think before making a tent!

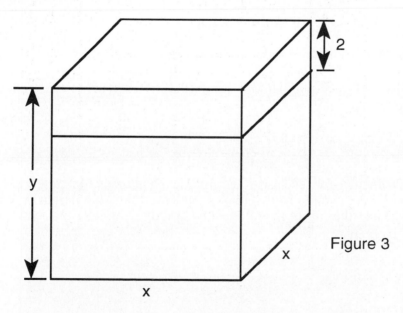

Figure 3

Project 9B
The following problems deal with alternative methods of constructing a hatbox.

1. Hattie the milliner needs a hatbox in the shape of a rectangular box with a square base and no top. The volume of the hatbox is to be 1000 in.^{3}, and the edge x of its base must be at least 8 in. long (Fig. 3). Hattie will also make a square lid with a 2-in. rim. Thus the hatbox-with-lid is, in effect, two open-topped boxes -- the first box with height y at least 2 in. and the lid with height 2 in. What should the dimensions x and y be to minimize the *total area* A of the two open-topped boxes?

Solve Hattie's problem. Begin by expressing the total area as a function A of x. Don't forget to find the domain of A. Show that the equation $A'(x) = 0$ simplifies to the cubic equation

$$x^3 + 2x^2 - 1000 = 0.$$

Instead of attempting to solve this equation exactly, graph $A(x)$ and $A'(x)$ on the same set of coordinate axes and zoom in on the zero of $A'(x)$.

You might like to use *Maple* to do some of the symbolic algebra in this problem. For instance, since the volume of the box is 1000 in.^{3}, its dimensions x and y must satisfy the volume

```
relation :=     (x^2)*y = 1000;
```

The total area A of the open box plus its lid is given by the

```
formula :=     A = x^2 + 4*x*y + x^2 + 8*x;
```

Now you can solve for A by eliminating the variable y:

```
subs( y = 1000/x^2, formula );
```

You should find that A is given by the function

```
A := x -> 2*x^2 + 8*x + 4000/x;
```

You will then discover the equation $x^3 + 2x^2 - 1000 = 0$ when you ask *Maple* to compute the equation

```
diff( A(x), x ) = 0;
```

2. Repeat Problem 1 for the case of a cylindrical hatbox as illustrated in Fig. 4. The circular base and lid are each to have radius r at least 4 in., the height h of the base must be at least 2 in., the rim of the lid must be exactly 2 in. tall, and the volume of the hatbox is to be 1000 in.^{3}

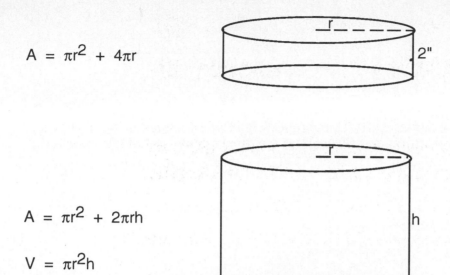

$$A = \pi r^2 + 4\pi r$$

$$A = \pi r^2 + 2\pi rh$$

$$V = \pi r^2 h$$

Figure 4

Project 10
Calculator/Computer Implementation of Newton's Method

Reference: Section 3.9 of Edwards & Penney

With calculators and computers that permit user-defined functions, Newton's method is very easy to set up and apply repeatedly. It is helpful to interpret Newton's iteration

$$x_{n+1} = x_n - \frac{f(x_n)}{f'(x_n)} \tag{1}$$

as follows. Having first defined the functions f and f', we then define the "iteration function"

$$g(x) = x - \frac{f(x)}{f'(x)}. \tag{2}$$

Newton's method is then equivalent to the following procedure. Begin with an initial estimate x_0 of the solution of the equation

$$f(x) = 0. \tag{3}$$

Calculate successive approximations x_1, x_2, x_3, ..., to the exact solution by means of the iteration

$$x_{n+1} = g(x_n). \tag{4}$$

That is, apply the function g to each approximation to get the next one.

Example

With the function $f(x) = x^3 - 4x + 1$, for which $f(0) = 1$ and $f(1) = -2$, thereby indicating a zero in the interval [0, 1], the following *Maple* commands implement Newton's method in the form described above (starting with the initial guess $x_0 = 0.5$).

```
f := x -> x^3 - 4*x + 1;
g := x -> x - f(x)/D(f)(x);

x := 0.5;
x := g(x);
```

After the functions f and g have been defined and the initial guess entered, the final command `x := g(x)` is entered repeatedly to generate successive Newton approximations until subsequent ones agree to the desired accuracy:

```
x := 0.5;
                   x := 0.5

x := g(x);
                   x := 0.230769

x := g(x);
                   x := 0.254000

x := g(x);
                   x := 0.254102

x := g(x);
                   x := 0.254102
```

Thus we discover the solution $x = 0.254102$ (accurate to six decimal places) of the equation $x^3 - 4x + 1 = 0$.

Project 10A

Figure 1 shows a large cork ball of radius 1 ft floating in water. If its density is 1/4 that of water, then Archimedes' law of buoyancy implies that the ball floats in water with 1/4 of its total volume submerged. Because the volume of the cork ball is $4\pi/3$, the volume of the part of the ball beneath the waterline is $V = \pi/3$. The volume of a spherical segment of radius r and height $h = x$ (as in Fig. 1) is given by the formula

$$V = \frac{\pi x}{6}(3r^2 + x^2).$$

This formula was derived by Archimedes.

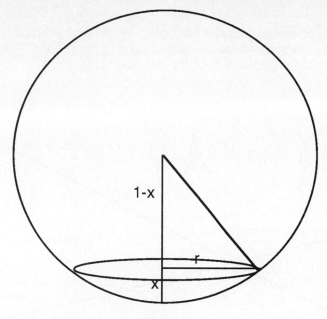

Figure 1

Proceed as follows to find the depth x to which the ball sinks in the water. Equate the two previous expressions for V, and then use the right triangle in Fig. 1 to eliminate r. You should find that x must be a solution of the cubic equation

$$f(x) = x^3 - 3x^2 + 1 = 0. \tag{5}$$

As the graph of f in Fig. 2 indicates, this equation has three real solutions -- one in $[-1, 0]$, one in $[0, 1]$, and one in $[2, 3]$. Obviously the solution between 0 and 1 gives the actual depth x to which the ball sinks. But use Newton's method to find all three solutions accurate to four decimal places.

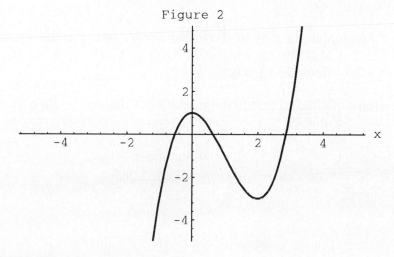

Figure 2

Project 10B
Investigate the cubic equation

$$4x^3 - 42x^2 - 19x - 28 = 0.$$

25

Perhaps you can see graphically that it has only a single real solution. Find it (accurate to four decimal places). First try the initial guess $x_0 = 0$; be prepared for at least 25 iterations. Then try initial guesses $x_0 = 10$ and $x_0 = 100$.

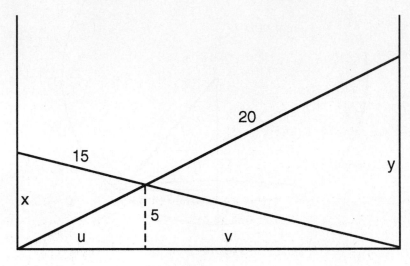

Figure 3

Project 10C

A 15-ft ladder and a 20-ft ladder lean in opposite directions against the vertical walls of a hall (Fig. 3). The ladders cross at a height of 5 ft. You must find the width w of the hall. First, let x and y denote the heights of the tops of the ladders on the walls, u and v the lengths shown in the figure, so that $w = u + v$. Use similar triangles to show that

$$x = 5\left(1 + \frac{u}{v}\right), \qquad y = 5\left(1 + \frac{v}{u}\right).$$

Then apply the Pythagorean theorem to show that $t = u/v$ satisfies the equation

$$t^4 + 2t^3 - 7t^2 - 2t - 1 = 0.$$

Finally, use Newton's method to find first the possible values of t, then those of w, accurate to four decimal places.

Chapter 4

Additional Applications of the Derivative

Project 11
Graphical Solution of Nonstandard Box Problems

Reference: Section 4.4 of Edwards & Penney

Figure 1 shows a rectangular box with square base. Suppose that its volume $V = x^2 y$ is to be 1000 in.^{3} We want to make this box at minimal cost. Each of the six faces costs a cents/in.^{2}, and gluing each of the 12 edges costs b cents per inch of edge length.

Figure 1

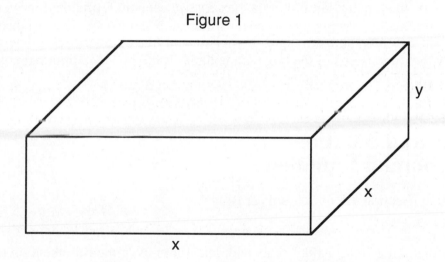

1.   Show first that the total cost C is given as a function of the base edge length x by

$$C(x) = 2ax^2 + 8bx + \frac{4000a}{x} + \frac{4000b}{x^2}. \qquad (1)$$

You might like to use *Maple* to eliminate the variable y between the volume

```
relation :=      (x^2)*y = 1000;
```

and the cost

```
formula :=    C = 2*a*x^2 + 4*a*x*y + 8*b*x + 4*b*y;
```

27

2.   Suppose that $a = b = 1$. Find the dimensions of the box of minimal cost graphically by zooming in on the appropriate solution of $C'(x) = 0$.

3.   Repeat Problem 2 with a and b being the last two nonzero digits of your student I.D. number.

4.   After doing Problems 2 and 3, you should smell a rat. Is it possible that the shape of the optimal box is independent of the values of a and b? Show that the equation $C'(x) = 0$ simplifies to the fourth-degree equation

$$f(x) = ax^4 + 2bx^3 - 1000ax - 2000b = 0. \tag{2}$$

Then use the *Maple* command

```
solve( f(x) = 0, x );
```

to solve this equation (for x). You may also find it interesting to solve Eq. (2) by hand -- begin by factoring x^3 from the first two terms and 1000 from the last two.

5.   Suppose that -- instead of all six faces of the box in Fig. 1 having the same cost of a cents/in.^{2} -- the top and bottom of the box cost p cents/in.^{2}, and the four vertical sides cost q cents/in.^{2} (while the 12 edges still cost b cents/in.). For instance, let p, q, and b be the last three nonzero digits of your student I.D. number. Then determine graphically the dimensions of the box with volume 1000 in.^{3} and minimal cost.

Project 12
Graphs and Solutions
of Polynomial Equations

Reference: Section 4.5 of Edwards & Penney

Project 12A
First show that, on a "reasonable" scale with integral units of measurement on the y-axis, the graph of the polynomial

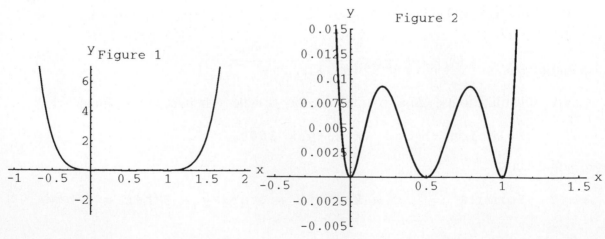

28

$$f(x) = \left[\frac{x}{6}(9x-5)(x-1)\right]^4 \qquad (1)$$

looks much like Fig. 1, with an apparent "flat section." Then produce a plot that reveals the true structure of the graph, as in Fig. 2. (Figures 1 and 2 are graphs of the equation $y = [x(x-1)(2x-1)]^2$ rather than the function in Eq. (1).) Finally, find (graphically or otherwise) the approximate coordinates of the local maximum and minimum points on the graph of $y = f(x)$. For instance, having defined the function

```
f := x -> ((x/6)*(9*x - 5)*(x - 1))^4;
```

you might use the *Maple* command

```
solve( D(f)(x) = 0, x );
```

to solve for its critical points.

Project 12B
The quartic equations

$$f(x) = x^4 - 55x^3 + 500x^2 + 11{,}000x - 110{,}000 = 0 \qquad (2)$$

and

$$g(x) = x^4 - 55x^3 + 550x^2 + 11{,}000x - 110{,}000 = 0 \qquad (3)$$

differ only in a single digit in the coefficient of x^2. However small this difference may seem, show graphically that Eq. (2) has four distinct real solutions but Eq. (3) has only two. Then find approximately the local maximum and minimum points on each graph.

Project 13
Finding Critical Points and Inflection Points on Exotic Graphs

Reference: Section 4.6 of Edwards & Penney

For each of Projects A and B below, choose in advance an integer n between 0 and 9. For instance, you could let n be the last digit of your student I.D. number.

Project 13A
If the coefficients a, b, and c are defined to be

$$a = 30{,}011 + 2n,$$
$$b = 30{,}022 + 4n, \quad \text{and}$$
$$c = 10{,}010 + 2n,$$

29

then the curve

$$y = 10,000x^3 - ax^2 + bx + c$$

has "two good wiggles like a good cubic should." Find them!

In particular, find the local maximum and minimum points and the inflection point (or points) on this curve. You can use the *Maple* command

```
solve( D(f)(x) = 0, x);
```

to locate the critical points and the command

```
solve( D(D(f))(x) = 0, x);
```

to locate the possible inflection points. Give the coordinates of each of these points accurate to five significant digits. Produce a graph that plainly exhibits all these points -- you can mark and label the points by hand. As you zoom in, you'll need to control carefully the successive viewing windows.

The *Maple* worksheet for this section contains an exposition of Project D below that illustrates efficient techniques for organizing computations of the coordinates of critical and inflection points. You may want to consult it before proceeding with Projects A, B, and C.

Project 13B
Your task is to analyze the structure of the curve

$$y = x^7 + 5x^6 - 11x^5 - 21x^4$$
$$+ 31x^3 - 57x^2 - (101 + 2n)x + (89 - 3n).$$

Provide the same information that was specified in Project A above. You will probably need to produce separate plots with different scales, showing different parts of this curve. In the end, use all the information accumulated to produce a careful hand sketch (not to scale) displaying all the maxima, minima, and inflection points on the curve with their (approximate) coordinates labeled.

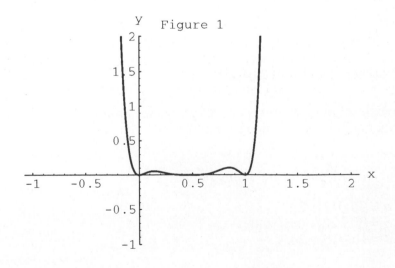

30

Project 13C

Let

$$f(x) = [x(x - 1)(2x - 1)(9x - 4)]^2.$$

The graph of f is shown in Fig. 1. Despite first appearances, this function has at least four local minima, three local maxima, and six inflection points in the interval $[0, 1]$. Find the approximate coordinates of all 13 points, and show the graph of f on a scale that makes all these points evident.

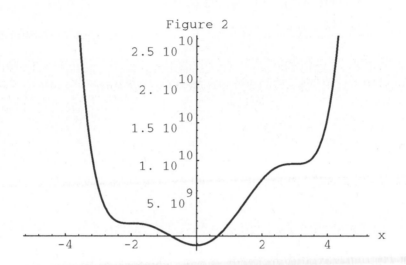

Figure 2

Project 13D

Explore in detail the structure of the graph of the function

$$f(x) = -1,234,567,890 + 2,695,140,459\ x^2 \\ + 605,435,400\ x^3 - 411,401,250\ x^4 \\ - 60,600,000\ x^5 + 25,000,000\ x^6$$

whose graph is shown in Fig. 2. (The *Maple* worksheet for this project tells where this exotic polynomial came from.) At a glance, it might appear that we have only three critical points -- a local minimum near the origin and two inflection points -- plus two more inflection points that are not critical points. Settle the matter. How many of each, in fact, are there? Find and exhibit all of them graphically.

31

Chapter 5

The Integral

Project 14
Numerical Calculation
of Riemann Sums

Reference: Section 5.4 of Edwards & Penney

In this project you will use *Maple* to calculate Riemann sums approximating the integral $\int_a^b f(x)\,dx$. Recall that a Riemann sum corresponding to a subdivision of $[a, b]$ into n equal subintervals of length $h = (b - a)/n = \Delta x$ is of the form

$$\sum_{i=1}^n f(z_i)\,\Delta x,$$

where the point z_i is selected in the ith subinterval $[x_{i-1}, x_i]$ for each $i = 1, 2, \ldots, n$. We get

- The **left-hand sum** L_n with $z_i = x_{i-1}$, the left-hand endpoint of $[x_{i-1}, x_i]$;

- The **right-hand sum** R_n with $z_i = x_i$, the right-hand endpoint of $[x_{i-1}, x_i]$;

- The **midpoint sum** M_n with $z_i = m_i = (x_{i-1} + x_i)/2$, the midpoint of $[x_{i-1}, x_i]$.

Thus

$$L_n = h \sum_{i=1}^n f(x_{i-1}), \qquad R_n = h \sum_{i=1}^n f(x_i), \qquad M_n = h \sum_{i=1}^n f(m_i). \qquad (1)$$

Figure 1 illustrates (with $f(x) = x^2$ and $n = 6$) the rectangular polygons whose areas are calculated by the sums in Eqs. (1).

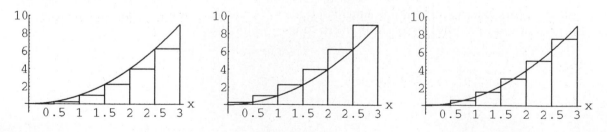

Figure 1

The *Maple* **sum** function is readily used to calculate the Riemann sums in Eqs. (1). The command

```
sum( g(i), i = p..q );
```

adds the values of the function g at the integer points $p, p + 1, p + 2, ...$, ending at the last such point that does not exceed q. Having defined

```
f := x ->x^2;              # for instance
a := 0;          b := 1;
n := 10;
h := (b - a)/n;
```

we can calculate the three sums in Eqs. (1) with the commands

```
leftSum  := evalf( h*sum(f(a + i*h),     i = 0..n-1));
rightSum := evalf( h*sum(f(a + i*h),     i = 1..n  ));
midSum   := evalf( h*sum(f(a - h/2 + i*h), i = 1..n ));
```

Do not proceed until you're sure that (and why) these sums do, indeed, give the left-hand, right-hand, and midpoint sums, respectively.

Project 14A

To check that your Riemann sum functions are working properly, first let

k = the largest digit in you student I.D. number

and then approximate the integral from $x = 0$ to $x = 1$ of the power function

```
f := x -> x^k;
```

Start by calculating each of your sums with $n = 10$ subintervals, and then double the number of subintervals successively, recording your results rounded off to three decimal places. Fill in the following table:

n	left sum	right sum	midpoint sum
10			
20			
40			
80			

Which of the three approximating sums appears to give the most accurate results (as compared with the exact value of the integral)?

Project 14B

Use both midpoint sums and the average

```
averageSum := (leftSum + rightSum)/2;
```

of the left-hand and right-hand sums to corroborate the claim that $\int_0^\pi \sin x \, dx = 2$.

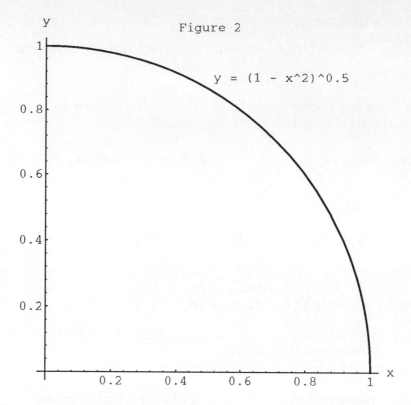

y

Figure 2

1

$y = (1 - x^2)^{0.5}$

0.8

0.6

0.4

0.2

0.2   0.4   0.6   0.8   1   x

Project 14C

First explain why Fig. 2 and the circle area formula $A = \pi r^2$ imply that

$$\int_0^1 4\sqrt{1-x^2}\ dx = \pi.$$

Then use midpoint-sum approximations to this sum to approximate the number π. Begin with $n = 10$ subintervals, then successively double n. Continue until two successive approximations corroborate the familiar approximation $\pi \approx 3.14 \approx 22/7$.

Project 15
Automatic Computation of Areas

Reference: Section 5.8 of Edwards & Penney

The *Maple* command

```
evalf( Int( f(x), x = a..b ) );
```

can be used to evaluate numerically the integral of the function $f(x)$ from $x = a$ to $x = b$. Thus the approximation

```
f := x -> x^2;
evalf( Int( f(x), x = 0..1 ) );
                0.3333333333
```

corresponds to the actual value 1/3.

34

For your very own versions of Projects A, B, and C below, choose a fixed integer *k* from 1 to 9 to use throughout. For instance, *k* could be the last nonzero digit of your student I.D. number.

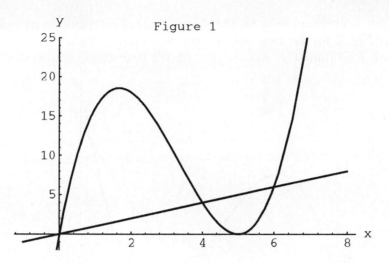

Figure 1

Project 15A

Let the two functions *f* and *g* be defined as follows:

$$f(x) = x, \tag{1}$$

$$g(x) = x(x - n - 1)^2. \tag{2}$$

As illustrated in Fig. 1 (for the case *n* = 4), the graphs $y = f(x)$ and $y = g(x)$ bound two regions R_1 and R_2. You are to find the sum *A* of the areas A_1 and A_2 of these two regions.

First solve manually (that is, by ordinary pencil-and-paper computation) for the *x*-coordinates of the three indicated points of intersection of $y = f(x)$ and $y = g(x)$. Then use your calculator or computer to evaluate the two integrals required to find *A* numerically. Also, calculate A_1 and A_2 exactly (by using the fundamental theorem of calculus), and see if your results agree.

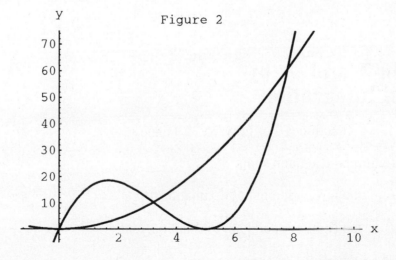

Figure 2

Project 15B
Repeat Project A, but replace the function $f(x)$ in Eq. (1) with

$$f(x) = x^2. \tag{3}$$

You can, as before, solve manually for the three points of intersection of $y = f(x)$ and $y = g(x)$ (shown in Fig. 2 for the case $n = 4$), but you surely would not want to calculate the areas A_1 and A_2 manually. Instead, use *Maple* to evaluate the necessary integrals numerically.

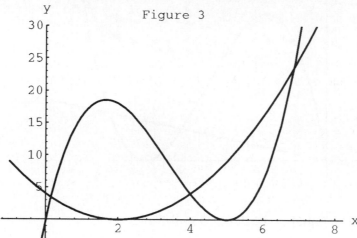

Project 15C
Repeat Project A, but replace the function $f(x)$ in Eq. (1) with

$$f(x) = \left(x - \frac{n}{2} \right)^2 \tag{4}$$

Once again a plot (Fig. 3 illustrates the case $n = 4$) shows two regions R_1 and R_2 bounded by $y = f(x)$ and $y = g(x)$. But now you would *not* be able to solve manually for the three points of intersection that provide the needed limits for the area integrals. Instead, use **fsolve** to find the points of intersection of the two graphs. Finally, find numerically the value of the areas A_1, A_2, and their sum A.

Project 16
Finding ln 2 and π by Numerical Integration

Reference: Section 5.9 of Edwards & Penney

Project 14 dealt with the Riemann sums

L_n , the left-endpoint approximation,

R_n , the right-endpoint approximation, and

M_n , the midpoint approximation

36

based on a subdivision of $[a, b]$ into n equal-length subintervals, to approximate the integral $\int_a^b f(x)\, dx$.

The Riemann sums L_n, R_n, and M_n suffice in turn to calculate the trapezoidal and Simpson approximations of this section. In particular, the trapezoidal approximation is given by the average

$$T_n = \frac{1}{2}\left(L_n + R_n\right) \tag{1}$$

of the left-endpoint and right-endpoint approximations. If n is an *even* integer, then Simpson's approximation with n equal-length subintervals is given by the *weighted* average

$$S_n = \frac{1}{3}\left(T_{n/2} + 2M_{n/2}\right) \tag{2}$$

of the midpoint and trapezoidal approximations with $n/2$ subintervals. The following *Maple* commands implement Eqs. (1) and (2):

```
f := x ->x^2;              # for instance

a := 0;    b := 1;

n := 5;                    # for Simpson's sum
h := (b - a)/n;            # with  2n  subintervals

leftSum  := evalf( h*sum(f(a + i*h), i = 0..n-1)):
rightSum := evalf( h*sum(f(a + i*h), i = 1..n  )):

trapSum  := (leftSum + rightSum)/2:

midSum   :=  evalf( h*sum(f(a - h/2 + i*h), i = 1..n )):

SimpSum  = (trapSum + 2*midSum)/3;
```

Important: Note that, because of the subscripts in (2), these commands with `n = 5` entered yield the value of Simpson's approximation with **10** (not 5) subintervals.

Project 16A
The natural logarithm (corresponding to the LN or, in some cases, the LOG key on your calculator) of the number 2 is the value of the integral

$$\ln 2 = \int_1^2 \frac{dx}{x}.$$

The value of $\ln 2$ correct to 15 decimal places is

$$\ln 2 \approx 0.69314\ 71805\ 59945.$$

37

See how many correct decimal places you can obtain in a reasonable period of time by using a Simpson's approximation procedure.

Project 16B
In Chapter 8 we will study the inverse tangent function $y = \arctan x$ (y is the angle between $-\pi/2$ and $\pi/2$ such that $\tan y = x$). There we will discover that the derivative of $y = \arctan x$ is given by

$$\frac{dy}{dx} = \frac{1}{1 + x^2}.$$

The fundamental theorem of calculus therefore implies that

$$\int_0^1 \frac{dx}{1 + x^2} = \left[\tan^{-1} x\right]_0^1 = \tan^{-1} 1 - \tan^{-1} 0 = \frac{\pi}{4}.$$

It follows that the number π is the value of the integral

$$\pi = \int_0^1 \frac{4}{1 + x^2}\, dx.$$

The value of π accurate to 15 decimal places is

$$\pi \approx 3.14159\ 26535\ 89793.$$

See how many correct decimal places you can obtain in a reasonable period of time by using a Simpson's approximation procedure.

Chapter 6

Applications of the Integral

Project 17
Numerical Approximation
of Volumes of Revolution

Reference: Section 6.2 of Edwards & Penney

The volume formula

$$V = \int_a^b \pi [f(x)]^2 \, dx \tag{1}$$

for volume of revolution by cross sections can be derived by using circular cylinders to approximate the volume of the solid obtained by revolving around the x-axis the region that lies under the curve $y = f(x)$ over the interval $[a, b]$. For instance, in the right-endpoint approximation

$$R_n = \sum_{i=1}^n \pi \left[f(x_i) \right]^2 \Delta x \tag{2}$$

with $\Delta x = (b - a)/n$, the ith term $\pi [f(x_i)]^2 \Delta x$ is the volume of the cylinder with radius $r = f(x_i)$ and height $h = \Delta x$ that approximates the slice of the solid corresponding to the subinterval $[x_{i-1}, x_i]$. The corresponding *Maple* command is

```
rightSum := h*evalf( Pi*sum(f(a + i*h)^2, i = 1..n  ));
```

using the notation of the previous three projects.

In the third century B.C. Archimedes regarded the sphere of radius r as a solid of revolution in deriving his famous volume formula $V = \dfrac{4}{3} \pi r^3$. A major difference is that his derivation used conical frusta rather than cylinders. Figure 6.2.42 in the text shows the approximating solid obtained by revolving around the x-axis the polygonal arc $P_0 P_1 P_2 \ldots P_n$ (where P_i denotes the point $(x_i, f(x_i))$ on the curve $y = f(x)$). The approximating slice that corresponds to the ith subinterval $[x_{i-1}, x_i]$ is the *conical frustum* labeled in Fig. 6.2.42. It has base radii $R = f(x_{i-1})$ and $r = f(x_i)$ and height $h = \Delta x$. By using the volume formula

$$V = \frac{\pi}{3} (r^2 + rR + R^2) h$$

for the volume of such a conical frustum, we therefore get the "frustum approximation"

39

$$F_n = \sum_{i=1}^{n} \frac{\pi}{3} \left\{ [f(x_{i-1})]^2 + f(x_{i-1})f(x_i) + [f(x_i)]^2 \right\} \Delta x$$

to the volume integral in Eq. (1). The corresponding *Maple* command is

```
frustSum := (h/3)*evalf(Pi*sum(f(a+(i-1)*h)^2 +
            f(a+(i-1)*h)*f(a+i*h) + f(a+i*h)^2, i=1..n));
```

In the following problems use *Maple* commands (as in Projects 14 and 16) to calculate the right- and left-endpoint approximations R_n and L_n, the trapezoidal approximation $T_n = (R_n + L_n)/2$, and the frustum approximation F_n to the volume integral in Eq. (1). In each case compare the accuracy of these approximations with $n =$ 10, 20, ... subintervals.

1.   Let $f(x) = x$ on $[0, 1]$, so that $V = \pi/3$ (why?).

2.   Let $f(x) = \sqrt{1-x^2}$ on $[0, 1]$. Explain why V is then the volume $2\pi/3$ of the hemisphere of radius 1.

3.   Let $f(x) = \sin x$ on $[0, \pi]$. Use the identity $\sin^2 x = (1 - \cos 2x)/2$ to show that $V = \pi^2$.

4.   Let $f(x) = \sec x$ on $[0, \pi/4]$, so that $V = \pi$ (why?).

Project 18
Volume Integrals and Custom-Designed Jewelry

Reference: Section 6.3 of Edwards & Penney

This project deals with the custom-made gold wedding band pictured in Fig. 1. Its shape is obtained by revolving the region A shown in Fig. 2 around the vertical axis shown there. The resulting wedding band has

- inner radius R,
- minimum thickness T, and
- width W.

The curved boundary of the region A is an arc of a circle whose center lies on the axis of revolution. For a typical wedding band, R might be anywhere from 6 to 12 millimeters (mm), T might be 0.5 to 1.5 mm, and W might be 4 to 10 mm. If a customer asks the price of a wedding band with given dimensions R, T, and W, the jeweler must first calculate the volume of the desired band to determine how much gold will be required to make it.

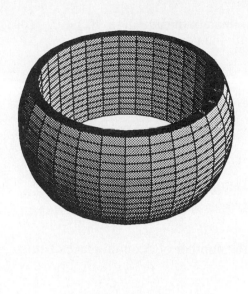

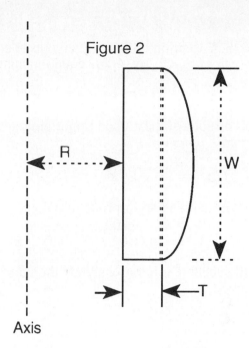

Figure 2

Axis

Figure 1

Problem 1   Use either the method of cross sections or the method of cylindrical shells to show that the volume V is given by the formula

$$V = \frac{\pi W}{6}(W^2 + 12RT + 6T^2)..$$ (1)

If these dimensions are measured in millimeters, then V is given in cubic millimeters. (There are 1000 mm³ in 1 cm³.)

       Suppose that the jeweler plans to charge the customer $1000 per *troy ounce* of alloy (90% gold, 10% silver) used to make the ring. (The profit on the sale, covering the jeweler's time and overhead in making the ring, is fairly substantial because the price of gold is generally under $400/oz and that of silver under $5/oz.) The inner radius R of the wedding band is determined by measurement of the customer's finger (in millimeters -- there are exactly 25.4 mm per inch). Suppose that the jeweler makes all wedding bands with $T = 1$ (mm). Then, for a given acceptable cost C (in dollars), the customer wants to know the maximum width W of the wedding band he or she can afford.

Problem 2   Measure your own ring finger to determine R (you can measure its circumference with a piece of string). Then choose a cost figure C in the $100 to $500 price range. Use Eq. (1) with $T = 1$ to find the width W of a band that costs C dollars (at $1000/oz, remember). You will need to know that the density of the gold-silver alloy is 18.4 gm/cm³ and that 1 lb contains 12 troy ounces and 453.59 gm. Use a handy computer to solve the resulting cubic equation in W.

       You should find that you need to solve a *cubic* equation in W. Recall that you can plot the graph $y = f(x)$ with the *Maple* command

41

```
plot( f(x), x =  a..b );
```

and then zoom in on the *x*-intercepts to solve approximately the equation $f(x) = 0$. You can ask *Maple* to solve this equation numerically with the command

```
fsolve( f(x) = 0, x );
```

This might also be a good opportunity to review **Newton's method**, which involves the *iteration function*

```
g := x -> x - f(x)/D(f)(x);
```

Starting with an appropriate initial guess **x0** you simply iterate the assignments

```
x1 := g(x0):          x0 := x1;
```

until successive iterations yield the same result (to the number of decimal places being retained).

Project 19
Numerical Approximation of Arc Length

Reference: Section 6.4 of Edwards & Penney

Arc-length integrals that appear in applications are seldom easy to evaluate exactly. This project involves the use of numerical integration techniques (as described in Projects 14 and 16) to approximate such integrals.

In Problems 1 through 10, calculate the right- and left-endpoint approximations R_n and L_n, the trapezoidal approximation $T_n = (R_n + L_n)/2$, and/or Simpson's approximation S_n. In each case, use $n = 10, 20, 40, \dots$ subintervals, doubling the number of subintervals until you are satisfied with the accuracy you obtain.

Find the arc length of each graph described in Problems 1-5.

1. $y = x^2, \quad 0 \le x \le 1$

2. $y = x^{5/2}, \quad 1 \le x \le 3$

3. $y = 2x^3 - 3x^2, \quad 0 \le x \le 2$

4. $y = x^{4/3}, \quad -1 \le x \le 1$

5. $y = 1 - x^2, \quad 0 \le x \le 100$

In Problems 6-10, find the areas of the surface of revolution generated by rotation of the given smooth arc around the given axis.

6. $y = x^2,$ $0 \le x \le 4;$ the *x*-axis

7. $y = x^2,$ $0 \le x \le 4;$ the *y*-axis

8. $y = x - x^2,$ $0 \le x \le 1;$ the *x*-axis

9. $y = x^2,$ $0 \le x \le 1;$ the line $y = 4$

10. $y = x^2,$ $0 \le x \le 1;$ the line $x = 2$

11.   Suppose that the main supporting cable for a suspension bridge has the shape of a parabola with equation $y = kx^2$. This cable joins the points $(-S, H)$ and $(+S, H)$ so the suspension bridge has total span $2S$, and the height of the cable (relative to its lowest point) is H at each end. Show that the total length of this cable is

$$L = 2 \int_0^S \left[1 + \frac{4H^2}{S^4} x^2 \right]^{1/2} dx.$$

12.   Italian engineers have proposed a single-span suspension bridge across the Strait of Messina (5 miles wide) between Italy and Sicily. The plans include suspension towers 1250 ft high at each end. Use the integral in Problem 11 to approximate the length L of the parabolic suspension cable for this proposed bridge. Assuming the given dimensions are exact, approximate the integral with sufficient accuracy to determine L to the nearest foot.

43

Chapter 7

Exponential and Logarithmic Functions

Project 20
Approximating the Number *e*
By Calculating Slopes

Reference: Section 7.1 of Edwards & Penney

Figure 1 shows the curve $y = a^x$ and its tangent line at the point $(0, 1)$. Applying the definition of the derivative, we find that the slope $m(a)$ is given by the limit

$$m(a) = \lim_{h \to 0} \frac{a^h - 1}{h}. \qquad (1)$$

Since the number e may be defined as the value of a such that the slope of the tangent line in Fig. 1 is precisely 1, we can investigate the value of e by numerical investigation of the limit in Eq. (1).

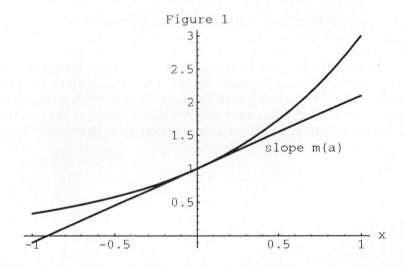

Figure 1

slope m(a)

For a fixed value of a, we can explore the limit in Eq. (1) by calculating the values of the fraction $p(h) = (a^h - 1)/h$ for a rapidly decreasing sequence h values such as $0.1, 0.01, 0.001, \dots, 10^{-n}$. For instance, with $a = 2$ the *Maple* commands

```
a := 3;
p := h -> (a^h - 1)/h;
for n from 1 to 7 do
    print(evalf(1/10^n),`    `,evalf(p(1/10^n))) od;
```

yield (with **Digits := 15**) the table

44

```
0.1              0.717735
0.01             0.695555
0.001            0.693387
0.0001           0.693171
0.00001          0.693150
0.000001         0.693147
0.0000001        0.693147
```

showing values of h in the first column and corresponding values of $p(h)$ in the second column. It is therefore apparent that the slope $m(2)$ of the tangent line to $y = 2^x$ at $(0, 1)$ is given by

$$m(2) \approx 0.693147 < 1,$$

If you re-execute the commands above with $a = 3$, you should find that

$$m(3) \approx 1.09861 > 1.$$

It therefore follows that the mysterious number e for which

$$m(e) = 1$$

lies somewhere between 2 and 3.

Your task in this project is to continue in this manner to bracket the number e more closely. For instance, interpolation between the values $m(2) \approx 0.693147$ and $m(3) \approx 1.09861$ above suggests that $e \approx 2.7$ or $e \approx 2.8$. Hence the next step is to investigate the numerical values of $m(2.7)$ and $m(2.8)$ to see whether it follows that $2.7 < e < 2.8$.

Continue in this way to "close in" on the number e. Don't quit until you're convinced that $e \approx 2.718$ accurate to three decimal places (rounded off). Record carefully the details of your investigation and the results along the way leading to your final conclusion.

Project 21
Approximating the Number e
By Numerical Integration

Reference: Section 7.2 of Edwards & Penney

Since $\ln e = 1$, the integral definition of the natural logarithm implies that

$$\int_1^e \frac{dx}{x} = 1. \tag{1}$$

This fact suggests the possibility of using numerical integration to investigate the value of e and thus to verify that

$$e \approx 2.71828.$$

The objective of this project is for you to "bracket" the number e between closer and closer numerical approximations -- attempting to determine the value of the upper limit b so that the value of the integral

$$\ln b = \int_1^b \frac{dx}{x} \tag{2}$$

is as close as possible to the target value 1 that we see in Eq. (1).

Start by applying the Simpson's approximation commands in Project 16, first with

```
f := x -> 1/x;          a := 1;     b := 2;     n := 25;
```

(thus actually 50 subintervals) to show that

$$\ln 2 \approx 0.693147,$$

and then with `b := 3` to show that

$$\ln 3 \approx 1.09861.$$

These results indicate that $b = 2$ is too small and that $b = 3$ is too large to yield 1 for the value of the integral in Eq. (2). Thus $2 < e < 3$.

Interpolation between 0.693147 and 1.09861 suggests that you try the upper limits $b = 2.6$, $b = 2.7$, and perhaps $b = 2.8$. The results you get when you do so should indicate that $2.7 < e < 2.8$.

As you bracket e more closely, you will need to increase the number of decimal places displayed and the number of subintervals used in Simpson's approximation to keep pace with your increasing accuracy. Continue until you have bracketed e between two seven-place approximations that both round to 2.71828.

Having carried out this numerical investigation faithfully using Simpson's approximation, you are perhaps entitled to try the "black box" approach of using *Maple*'s built-in numerical integration function. For instance, the results

```
evalf( Int( 1/x, x = 1..2 ) );
                    0.693147
```

```
evalf( Int( 1/x, x = 1..3 ) );
                    1.098610
```

again start us off with the estimate $2 < e < 3$.

Project 22
Approximating the Number *e*
By Successive Squaring

Reference: Section 7.3 of Edwards & Penney

In this brief project you are to use the limit

$$e = \lim_{n \to \infty} \left(1 + \frac{1}{n}\right)^n \tag{1}$$

to investigate (once again) the numerical value of the number e. Assuming that this limit exists, you can "accelerate" the convergence to the limit by calculating the quantity $(1 + 1/n)^n$ only for each power $n = 2^k$ of 2 instead of for every positive integer n. That is, consider the sequence $n = 1, 2, 4, 8, ..., 2^k, ...$ instead of the sequence $n = 1, 2, 3, 4, ...$ of all positive integers:

$$e = \lim_{k \to \infty} \left(1 + \frac{1}{2^k}\right)^{2^k}. \tag{2}$$

This approach has the advantage that the terms

$$s_k = \left(1 + \frac{1}{2^k}\right)^{2^k} \tag{3}$$

approaching e in Eq. (2) can be calculated by "successive squaring," because

$$(x^2)^2 = x^4, \quad (x^4)^2 = x^8, \quad (x^8)^2 = x^{16},$$

and so forth. This can be done readily with a simple calculator possessing only an x^2 or "squaring" key. Thus to calculate the 2^kth power $x^{\left(2^k\right)}$ of the number x, you need only enter x, then press the squaring key k times in succession. (On some calculators, such as the TI-81, you must first press the x^2 and then ENTER each time.)

Consequently, you can calculate the number s_k in Eq. (3) by the following elementary steps:

1. Calculate 2^k (no squaring here).
2. Calculate the reciprocal $1/2^k$.
3. Add 1 to get the sum $1 + (1/2^k)$.
4. Square the result k times in succession.

Do this with $k = 2, 4, 6, 8, ..., 18, 20$. Construct a table showing each result s_k accurate to five decimal places. (When you finish you will *really know* that $e \approx$ 2.71828. Congratulations!) In lieu of a little hand-held calculator, you can use *Maple* to make a big calculator out of your computer. For example, you might calculate the value of

$$s_{10} = \left(1 + \frac{1}{2^{10}}\right)^{2^{10}}$$

as follows: First we raise 2 to the 10th power by simple multiplication.

```
two10th := 2*2*2*2*2*2*2*2*2*2;      # 10 two's
                    1024
```

Then we take the reciprocal and add 1:

```
x := 1.0 + 1/two10th;
```
$$1.00098$$

Finally, we square x 10 times by entering the command $x = x^2$ 10 times in succession:

```
x := x^2;           #  (Press Enter 10 times)
```
$$2.71696$$

At this point we have obtained the approximation $e \approx 2.717$ without the use of any transcendental operations, just "simple arithmetic."

Project 23
Going Graphically Where No One Has Gone Before

Reference: Section 7.4 of Edwards & Penney

This project calls for you to investigate the equation

$$2^x = x^{10}. \tag{1}$$

1. The graphs of $y = 2^x$ and $y = x^{10}$ (Fig. 1) *suggest* that Eq. (1) has two solutions -- one positive and the other negative. Use *Maple* to find these two solutions (each accurate to three or four decimal places) by successive magnification (the method of "zooming").

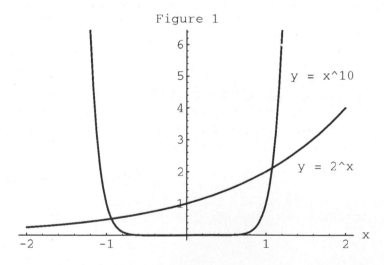

Figure 1

2. Figure 2 seems to indicate that x^{10} leaves 2^x forever behind as $x \to +\infty$. Show, however, that Eq. (1) has the same positive solutions as the equation

$$\frac{\ln x}{x} = \frac{\ln 2}{10}.$$

Hence conclude from the graph of $y = (\ln x)/x$ (Fig. 3) that Eq. (1) has precisely *two* positive solutions!

48

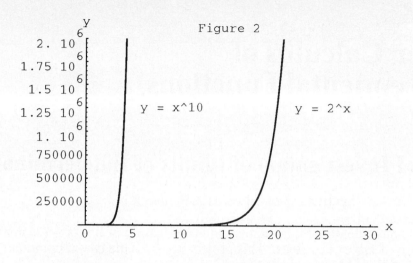

Figure 2

y = x^10   y = 2^x

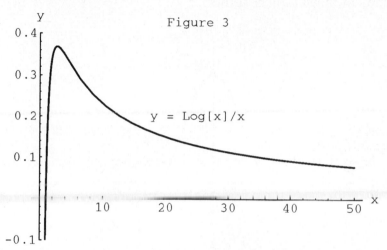

Figure 3

y = Log[x]/x

3.   Tabulate values of 2^x and x^{10} for $x = 10, 20, 30, 40, 50,$ and 60 and thereby verify that the missing positive solution is somewhere between $x = 50$ and $x = 60$. If you attempt to locate this solution by successive magnifications, you may well have the feeling of going boldly where no one has gone before!

4.   Use Newton's method to approximate (with four-place accuracy) all three solutions of Eq. (1).

49

Chapter 8

Further Calculus of Transcendental Functions

Project 24
Graphical Investigation of Limits of Indeterminate Forms

Reference: Section 8.3 of Edwards & Penney

Earlier projects have dealt with the numerical investigation of limits -- for instance, by the construction of tables of values. This project involves the use of computer graphing to investigate limits of indeterminate forms.

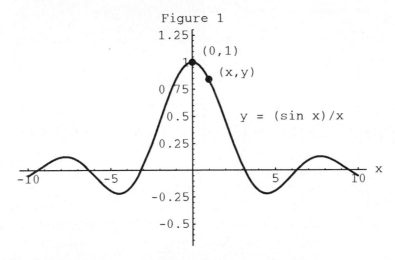

For example, consider the fundamental trigonometric limit. From the graph of $y = (\sin x)/x$ in Fig. 1 it seems clear that if the number x is close to zero, then the point (x, y) on the curve is near the point $(0, 1)$. This observation corroborates the fact that

$$\lim_{x \to 0} \frac{\sin x}{x} = 1. \tag{1}$$

We want to explore graphically the meaning of one number or point being "close" to another. Recall from Section 2.2 the meaning of the limit: We say that

$$\lim_{x \to a} f(x) = L \tag{2}$$

provided that, given any $\varepsilon > 0$, there exists a number $\delta > 0$ such that

$$0 < |x - a| < \delta \quad \text{implies that} \quad |f(x) - L| < \varepsilon. \tag{3}$$

The geometric meaning of the implication in (3) is this:

Suppose that two horizontal lines $y = L - \varepsilon$ and $y = L + \varepsilon$ above and below the point (a, L) are drawn. Then it is possible to draw two vertical lines $x = a - \delta$ and $x = a + \delta$ on either side of (a, L) such that the portion of the graph $y = f(x)$ (with $x \neq a$) that lies between the two vertical lines also lies between the two horizontal lines.

The question is this. Given a specific value of $\varepsilon > 0$, what is a value of $\delta > 0$ that makes this so? With *Maple* we can first plot the horizontal lines $y = L - \varepsilon$ and $y = L + \varepsilon$, then experiment with different locations of the vertical lines $x = a - \delta$ and $x = a + \delta$ to find a value of δ that "works" with the given value of ε.

Figure 2

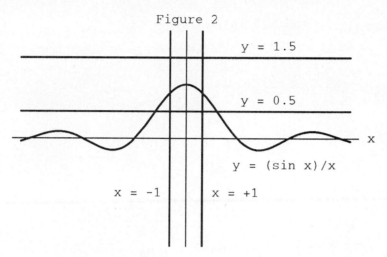

Thus, in Fig. 2 we see that every point on $y = (\sin x)/x$ between the vertical lines $x = -1$ and $x = +1$ also lies between the two horizontal lines $y = 0.5$ and $y = 1.5$. So for the limit in Eq. (1) -- for which $a = 0$ and $L = 1$ in (3) -- it follows that if $\varepsilon = 0.5$, then we can choose $\delta = 1$. The *Maple* worksheet for this project shows how the lines in Fig. 2 were placed (using the **display** function).

Figure 3

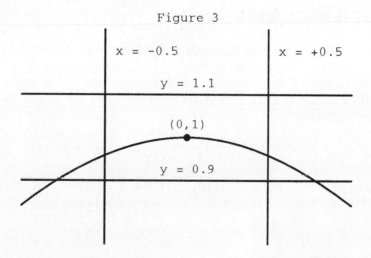

The smaller the value of $\varepsilon > 0$, the smaller a value of $\delta > 0$ we should expect to need. In Fig. 3 we see that if $\varepsilon = 0.1$, then we can choose $\delta = 0.5$.

51

In each of the following problems, a limit $\lim_{x \to a} f(x)$ is given. First inspect the graph $y = f(x)$ near $x = a$ to determine the apparent value L of the limit. Then experiment with horizontal and vertical lines to find values of $\delta > 0$ that satisfy the definition of the limit for the indicated values of $\varepsilon > 0$. Finally, apply l'Hôpital's rule to verify that $\lim_{x \to a} f(x) = L$.

1. $\lim_{x \to 0} \dfrac{e^x - 1}{x}$; $\qquad \varepsilon = 1, \ 0.5, \ 0.1$

2. $\lim_{x \to 0} \dfrac{1 - \cos x}{x^2}$; $\qquad \varepsilon = 0.25, \ 0.1, \ 0.05$

3. $\lim_{x \to 0} \dfrac{\sqrt{1 + x} - 1}{x}$; $\qquad \varepsilon = 0.25, \ 0.1, \ 0.05$

4. $\lim_{x \to 4} \dfrac{\sqrt{x} - 2}{x - 4}$; $\qquad \varepsilon = 0.1, \ 0.05, \ 0.01$

5. $\lim_{x \to 0} \dfrac{\tan x - \sin x}{x^3}$; $\qquad \varepsilon = 0.25, \ 0.1, \ 0.05$

6. $\lim_{x \to 0} \dfrac{\tan^{-1} x - \sin^{-1} x}{x^3}$; $\qquad \varepsilon = 0.25, \ 0.1, \ 0.05$

Project 25
Mathematics of the St. Louis Gateway Arch

Reference: $\qquad$ Section 8.5 of Edwards & Penney

Construction of the Gateway Arch in St. Louis, Missouri, was completed in October of 1965. With a hollow stainless steel design, its centroid curve (or center line) is described very closely by the equation

$$y = 693.86 - 68.767 \cosh \frac{3x}{299}. \tag{1}$$

Here y denotes height above the ground (in feet) and x denotes horizontal distance (in feet) from the arch's vertical axis of symmetry. You might begin by plotting Eq. (1) to verify that the centroid curve has the appearance of Fig. 1. Problems 1 through 5 involve mathematical aspects of the arch and its centroid curve.

1. $\qquad$ Deduce from Eq. (1) that the height H of the centroid curve's highest point is close to 625 ft and that the width W of the base of the arch is slightly less than 600 ft.

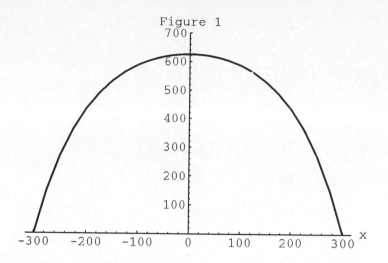

Figure 1

2.   Find the arc length of the centroid curve of the Gateway Arch. You may integrate numerically as in Projects 14 and 16.

The arch itself is a hollow tube with varying cross section. Each cross section normal to the centroid curve is an equilaterial triangle. The area (in square feet) of the cross section that intersects the centroid curve at the point (x, y) is

$$A = 1262.67 - (1.8198)y. \qquad (2)$$

3.   Show that the edge length E of the arch's triangular cross section varies from about 54 ft at its base to about 17 ft at its top.

4.   The perimeter of a triangular cross section with edge E is $P = 3E$. Explain heuristically the formula

$$s = \int_{\underline{\ }}^{\overline{\ }} P \, ds$$

that gives the surface area S of a tube in terms of its cross-sectional perimeter P and arc length s along its centroid curve. Then apply this formula appropriately to approximate numerically the (outer) surface area of the Gateway Arch. You might begin by expressing the perimeter as a function $P(x)$ and writing

$$ds = \frac{ds}{dx} \, dx = \sqrt{1 + \left(\frac{dy}{dx}\right)^2} \, dx.$$

5.   Explain heuristically the formula

$$V = \int_{\underline{\ }}^{\overline{\ }} A \, ds$$

that gives the volume V of a tube in terms of its cross-sectional area A and the arc length s along its centroid curve. Then apply this formula appropriately to find the total volume enclosed by the outer surface of the Gateway Arch.

Chapter 9

Techniques of Integration

Project 26
When Are Two Different Answers (for the Same Integral) Equivalent?

Reference: Section 9.2 of Edwards & Penney

According to a typical formula in a table of integrals,

$$\int \sqrt{x^2 + 1}\, dx = G(x) + C \tag{1}$$

where

$$G(x) = \frac{x}{2}\sqrt{x^2 + 1} + \frac{1}{2}\ln\!\left(x + \sqrt{x^2 + 1}\right). \tag{2}$$

But according to Serge Lang's *First Course in Calculus* (5th edition, Springer-Verlag, 1991, p. 376), this same indefinite integral is given by

$$\int \sqrt{x^2 + 1}\, dx = H(x) + C \tag{3}$$

where

$$H(x) = \frac{1}{8}\left[\left(x + \sqrt{x^2 + 1}\right)^2 + 4\ln\!\left(x + \sqrt{x^2 + 1}\right) - \left(x + \sqrt{x^2 + 1}\right)^{-2}\right]. \tag{4}$$

Your mission in this project is to determine whether the functions $G(x)$ and $H(x)$ in Eqs. (2) and (4) are, in fact, both antiderivatives of the same function

$$f(x) = \sqrt{x^2 + 1}. \tag{5}$$

We list below possible ways to investigate the relationship between the functions $f(x)$, $G(x)$, and $H(x)$. You should explore several different avenues of approach.

1.   Use *Maple* to plot both $y = G(x)$ and $y = H(x)$. If the functions $G(x)$ and $H(x)$ are both antiderivatives of $f(x)$, how should their graphs be related?

2.   Plot both the graph of the function $f(x)$ in (5) and the *derivatives* $G'(x)$ and $H'(x)$ of the functions in (2) and (4). Does the visual evidence convince you that

$$G'(x) = H'(x) = f(x)\,?$$

3.   Plot $f(x)$ and the quotients

$$\frac{G(x+h) - G(x-h)}{2h} \quad \text{and} \quad \frac{H(x+h) - H(x-h)}{2h},$$

which -- with $h = 0.001$ -- should closely approximate the derivatives $G'(x)$ and $H'(x)$, respectively. (Why?)

4.　By numerical integration -- using *Maple*'s `evalf(Int(f(x), x=a..b ))` function, for instance -- you can determine whether

$$\int_a^b f(x)\, dx = G(b) - G(a) = H(b) - H(a) \tag{6}$$

as the fundamental theorem implies if $G' = H' = f$. It should be fairly convincing if you can verify Eq. (6) numerically with several different pairs of limits, such as $a = 1,\ b = 5$ and $a = 7,\ b = 11$.

5.　Even without heavy metal hardware, you can compute numerical values of $G(x)$, $H(x)$, $G'(x)$, $H'(x)$, and $f(x)$ for several selected values of x. Do the numerical results imply that $G(x) = H(x)$ or that $G'(x) = H'(x) = f(x)$?

6.　But with *Maple* it's a simple matter to calculate the derivatives $G'(x)$ and $H'(x)$ symbolically to determine whether or not both are equal to $f(x)$. Perhaps you can do this even the old-fashioned way -- by using nothing but paper and pencil. And you might investigate in the same way whether or not $G(x) = H(x)$.

Project 27
Bounded Population Growth and the Logistic Equation

Reference:　　　　Section 9.5 of Edwards & Penney

The **logistic equation** is a differential equation of the form

$$\frac{dP}{dt} = k\, P\, (M - P) \qquad (k,\ M \text{ constants}). \tag{1}$$

The logistic equation models many animal (including human) populations more accurately than does the natural growth equation $dP/dt = kP$ for the population $P(t)$. For instance, think of an environment that can support a population of at most M individuals. We might then think of $M - P$ as the potential for further expansion when the population is P. The hypothesis that the rate of change dP/dt is therefore proportional to $M - P$ as well as to P itself then yields Eq. (1) with the proportionality constant k. The classic example of such a limited-environment situation is a fruit fly population in a closed container.

　　　The object of this project is to investigate the behavior of populations that can be modeled by the logistic equation.

1.　First separate the variables in Eq. (1) and then integrate using partial fractions to derive the solution

$$P(t) = \frac{MP_0}{P_0 + (M - P_0)e^{-kMt}} \tag{2}$$

that satisfies the initial condition $P(0) = P_0$. If k and M are positive constants, then

$$\lim_{t \to \infty} P(t) = M. \tag{3}$$

Hence the significance of the constant M in the logistic equation is that it is the *limiting population* that $P(t)$ approaches as t increases without bound.

2.   During the period from 1790 to 1930, the U.S. population $P(t)$ (t in years) grew from 3.9 million to 123.2 million. Throughout this period $P(t)$ remained close to the solution of the initial value problem

$$\frac{dP}{dt} = 0.03135\,P - 0.0001589\,P^2, \qquad P(0) = 3.9.$$

Has this differential equation continued since 1930 to predict accurately the U.S. population? If so, what is the limiting population of the United States?

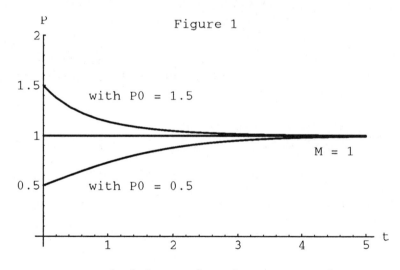

3.   For your very own logistic equation, choose nonzero integers r and s (for example, the last two nonzero digits of your student I.D. number), and then take

$$M = 10r, \qquad k = \frac{s}{100M}$$

in Eq. (1). Then plot the corresponding solution in Eq. (2) with several different values of the initial population P_0. What determines whether the graph $P = P(t)$ looks like the upper curve or the lower curve in Fig. 1?

4.   Now, with the same fixed value of the limiting population M as in Problem 3, plot solution curves with both larger and smaller values of k. What appears to be the relationship between the size of k and the rate at which the solution curve approaches its horizontal asymptote $P = M$?

The following sections provide an optional introduction to the powerful differential equations solving capabilities of *Maple*.

Symbolic Solution of Differential Equations

The following commands illustrate how to ask *Maple* to solve the logistic equation symbolically (without our having to separate the variables and do the integration ourselves). First we define the differential equation to be solved:

```
k := 'k';       M := 'M';       #  k and M now symbolic
P := 'P';
logisticDE := diff(P(t),t) = k*P(t)*(M - P(t));
```

Then we use the *Maple* `dsolve` function, including the initial condition `P(0) = P0` along with the logistic differential equation to be solved:

```
dsolve( {logisticDE, P(0) = P0}, P(t) );
```

You will want to check to see that this gives the same solution as that given in Eq. (2) above.

Numerical Solution of Differential Equations

There are numerical methods for the approximate solution of differential equations -- analogous to numerical methods for the approximation of definite integrals. The following commands illustrate the use of *Maple*'s `dsolve` function with the **numeric** option to solve the logistic equation numerically:

```
k := 1;    M := 1;    P0 := 0.5;
logisticDE := diff(P(t),t) = k*P(t)*(M - P(t));
solution :=
      dsolve( {logisticDE,P(0) = P0}, P(t), numeric );
```

You can now define and plot this numerical solution:

```
f := t -> solution(t)[2];
plot( f, t = 0..5, P = -0.1..2 );
```

This gives the lower solution curve in Fig. 1. With $P0 = 1.5$ we would have gotten the upper curve. (Try it!)

Project 28
Numerical Approximation of Improper Integrals

Reference: Section 9.8 of Edwards & Penney

The improper integral $\int_0^\infty e^{-x^2} dx$ is important in applications ranging from probability and statistics (political polling, for example) to traffic flow and the theory of heat.

Because the function $f(x) = \exp(-x^2)$ has no elementary antiderivative as a finite combination of familiar functions, a simple and direct evaluation of the limit

$$\int_0^\infty e^{-x^2}\, dx = \lim_{b\to\infty}\int_0^b e^{-x^2}\, dx \tag{1}$$

by using only the fundamental theorem of calculus is not feasible. But the fact that $e^{-x^2} \le e^{-x}$ for $x \ge 1$ implies that the improper integral in Eq. (1) converges rather than diverges to infinity. (Can you explain why?)

In the chapter on multiple integrals we will see that the exact value of the integral in Eq. (1) is given by

$$\int_0^\infty e^{-x^2}\, dx = \frac{\sqrt{\pi}}{2} \approx 0.886227. \tag{2}$$

To verify this value numerically, use the *Maple* command

```
evalf( Int( exp(-x*x), x = 0..b ) );
```

to compute values of the integral $\int_0^b e^{-x^2}\, dx$ with successively larger values of the upper limit b, such as $b = 1, 2, 3, ..., 10$. You might even assemble your results in a convincing `table`.

If using the automatic command above makes you feel guilty, apply any of the numerical integration methods described in Projects 14 and 16. If you use a method such as Simpson's approximation, it's good to increase the accuracy by, say, quadrupling the number of subintervals with each doubling of the value b of the upper limit.

Use one of the techniques indicated here to verify numerically the values specified in Problems 1 through 5.

1. $\displaystyle\int_0^\infty x^5 e^{-x}\, dx = 120$

2. $\displaystyle\int_0^\infty \frac{\sin x}{x}\, dx = \frac{\pi}{2}$

3. $\displaystyle\int_0^\infty \frac{dx}{x^2+1} = \frac{\pi}{2\sqrt{2}}$

4. $\displaystyle\int_0^\infty e^{-x^2}\cos 2x\, dx = \frac{\pi}{2e}$

5. $\displaystyle\int_0^\infty \frac{1-e^{-3x}}{x}\, dx = \frac{\ln 10}{2}$

Chapter 10

Polar Coordinates and Conic Sections

Project 29
Calculator/Computer-Generated Polar-Coordinates Graphs

Reference: Section 10.2 of Edwards & Penney

The graph of the polar-coordinates equation $r = f(\theta)$ can be plotted in rectangular coordinates by using the equations

$$x = r\cos\theta = f(\theta)\cos\theta \tag{1}$$

$$y = r\sin\theta = f(\theta)\sin\theta. \tag{2}$$

Then, as θ ranges from 0 to 2π (or, in some cases, through a much larger domain), the point (x, y) traces the polar graph $r = f(\theta)$.

Figure 1

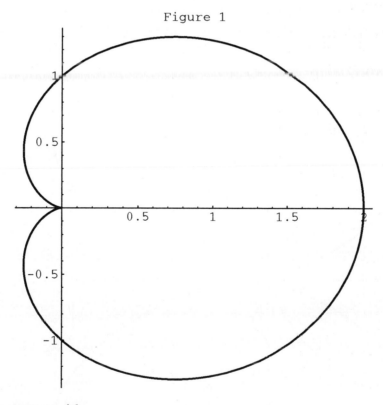

For instance, with

$$r = 1 + \cos\theta , \tag{3}$$

the equations

$$x = (1 + \cos\theta)\cos\theta$$

$$y = (1 + \cos\theta)\sin\theta$$

yield the cardioid shown in Fig. 1.   To graph this cardioid with the parametric version of the *Maple* `plot` command, we first enter the definitions

```
r := t -> 1 + cos(t);          #  Equation (3)
x := t -> r(t)*cos(t);         #  Equation (1)
y := t -> r(t)*sin(t);         #  Equation (2)
```

using `t` in place of θ. Then we graph the cardioid with the command

```
plot( [x(t), y(t), t = 0..2*Pi],
            x = -0.5..2.5, y = -1.5..1.5 );
```

Project 29A
Plot the polar-coordinates curve

$$r = (a + b \cos m\theta)(c + d \sin n\theta)$$

with various values of the coefficients a, b, c, d, and the positive integers m and n. You might begin with the special case $a = 1, b = 0$, or the special case $c = 1, d = 0$, or the special case $a = c = 0$.

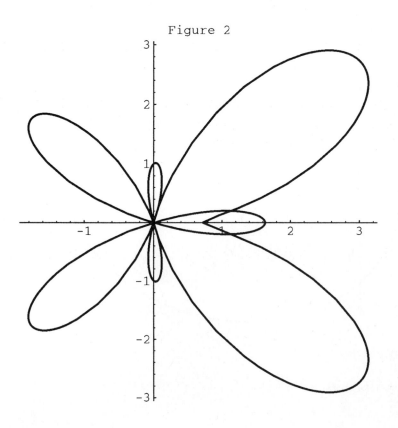

Figure 2

Project 29B

The simple "butterfly" shown in Fig. 2 is the graph of the polar-coordinates equation

$$r = e^{\cos\theta} - 2\cos 4\theta.$$

Now plot the polar-coordinates equation

$$r = e^{\cos\theta} - 2\cos 4\theta + \sin^5\frac{\theta}{12}$$

for $0 \le \theta \le 24\pi$. The incredibly beautiful curve that results was discovered by Temple H. Fay. His article "The Butterfly Curve" (*American Mathematical Monthly*, May 1989, p. 442) is well worth a trip to the library.

Project 30
Numerical Approximation of Polar-Coordinates Areas

Reference: Section 10.3 of Edwards & Penney

We can apply the numerical integration techniques of Projects 14 and 16 to the polar-coordinate area integral

$$A = \int_{\alpha}^{\beta} \frac{1}{2}[f(\theta)]^2 \, d\theta \tag{1}$$

to approximate numerically the area of the region bounded by the curve $r = f(\theta)$, $\alpha \le \theta \le \beta$.

For instance, let t_i denote the *midpoint* of the ith subinterval $[\theta_{i-1}, \theta_i]$ of a subdivision of $[\alpha, \beta]$ into n subintervals each of length $\Delta\theta = (\beta - \alpha)/n$. Then the corresponding midpoint approximation to the area in Eq. (1) is

$$A_n = \sum_{i=1}^{n} \frac{1}{2}[f(t_i)]^2 \Delta\theta. \tag{2}$$

Example
To approximate the area bounded by the cardioid $r = 1 + \cos\theta$, $0 \le \theta \le 2\pi$, we can use the *Maple* commands

```
r := t-> 1 + cos(t);

g := t -> (1/2)*r(t)^2;        #   integrand in (1)
a := 0;    b := 2*Pi;

n := 10;
h := (b - a)/n;
midSum := evalf( h*sum(g(a - h/2 + i*h), i = 1..n ) );
```

61

With both $n = 10$ and $n = 20$ this gives $A = 4.71239$ in agreement with the exact computation

```
int( (1/2)*(1 + cos(t)^2), t = 0..2*Pi );
```

$$3/2 \ Pi$$

In the following problems, first approximate the area of the given region R by using Eq. (2) with $n = 10$ and with $n = 20$. Then compare your approximations with the exact area calculated using Eq. (1).

1. R is the unit circle bounded by $r = 1$. Do you get the same results with $n = 10$ as with $n = 20$? Why?

2. R is the unit circle with boundary $r = 2\cos\theta$.

3. R is bounded by the cardioid $r = 2 - 2\sin\theta$.

4. R is bounded by the limaçon $r = 3 + 2\sin\theta$.

5. R is bounded by one loop of the three-leaved rose $r = 3\sin 3\theta$.

6. R is bounded by one loop of the eight-leaved rose $r = 2\cos 4\theta$ (Fig. 1).

7. R is bounded by one loop of the curve $r^2 = 4\cos 2\theta$.

Figure 1

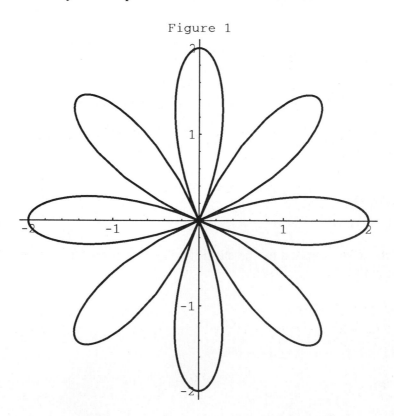

Chapter 11

Infinite Series

Project 31
Numerical Summation
of Infinite Series

Reference: Section 11.3 of Edwards & Penney

A computer (or programmable calculator) is indispensable for the calculation of partial sums with large numbers of terms.   Given the convergent infinite series

$$\sum_{n=1}^{\infty} a_n = a_1 + a_2 + a_3 + \ldots + a_n + \ldots, \tag{1}$$

we can use the *Maple* **sum** function to calculate the kth partial sum

$$S_k = \sum_{n=1}^{k} a_n = a_1 + a_2 + \ldots + a_k. \tag{2}$$

We need only define first a formula such as

```
a := n -> 1/3^n;                                    (3)
```

for the geometric series $\sum 1/3^n$ or

```
a := n -> 1/n^3;                                    (4)
```

for the series $\sum 1/n^3$,   giving the nth term $a_n = a(n)$ in (1) as a function of n. Then the numerical value of the kth partial sum S_k is given by

```
k = 25;          #  for instance
evalf( sum( a(n), n = 1..k ) );
```

In the case of an infinite series $\sum_{n=0}^{\infty} a_n$ beginning at $n = 0,$ we use instead the index range **n = 0..k**.

Example
With the term definition in (4) the command **evalf(sum(a(n), n = 1..k))** yields the following results:

k	25	50	100	200	300
S_k	1.20129	1.20186	1.20201	1.20204	1.20205

The exact value (in closed form) of the sum of the infinite series $\sum 1/n^3$ is not known, but it appears that

$$\sum_{n=1}^{\infty} \frac{1}{n^3} \approx 1.202$$

accurate to three decimal places, if we accept the rule of thumb -- commonly used but not always reliable -- of carrying two additional decimal places beyond the accuracy desired.

Project A

Calculate partial sums of the geometric series $\sum_{n=0}^{\infty} r^n$ with $r = 0.2$, 0.5, 0.75, 0.9, 0.95, and 0.99. For each value of r, calculate k-term partial sums with $k = 10$, 20, 30, ..., continuing until two successive results agree to four or five decimal places. (For $r = 0.95$ and 0.99, you may decide to use $k = 100$, $200, 300$, ...) How does the apparent rate of convergence -- as measured by the number of terms required for the desired accuracy -- depend on the value of r?

Project B

Calculate partial sums of the harmonic series $\sum_{n=1}^{\infty} 1/n$ with $k = 100$, 200, 300, ... terms (or with $k = 1000$, 2000, 3000, ... if you have an especially powerful microcomputer). Interpret your results in light of the known divergence of the harmonic series.

Project C
Euler originally introduced the famous number e as the sum of the infinite series

$$e = \sum_{n=0}^{\infty} \frac{1}{n!} = 1 + \sum_{n=1}^{\infty} \frac{1}{n!} \qquad\qquad \text{(with } 0! = 1),$$

where the *factorial* $n! = (1)(2)(3) \dots (n)$ denotes the product of the first n positive integers. Sum enough terms of this rapidly convergent infinite series to convince yourself that $e = 2.71828\ 1828$ (accurate to nine decimal places).

Project 32
Graphs of Taylor Polynomials

Reference: Section 11.4 of Edwards & Penney

By plotting several successive Taylor polynomials of a function on the same set of coordinate axes, one gets a visual sense of the way in which the function is approximated by partial sums of its Taylor series. For instance, Fig. 1 shows the graph of the function $f(x) = \sin x$ and its Taylor polynomials of degrees $n = 3, 5, 7, \dots, 17$.

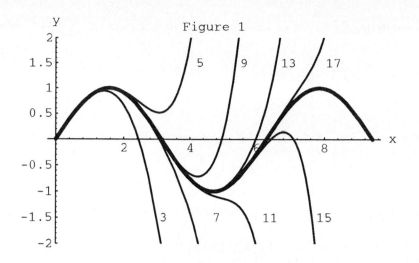

Figure 1

The *Maple* function

```
series( f(x), x = a, n );
```

yields the Taylor series of the function $f(x)$, centered at the point $x = a$, through the term x^{n-1} of degree $n - 1$. For instance,

```
S := series( sin(x), x = 0, 6 );
```

$$S := x - 1/6\ x^3 + 1/120\ x^5 + O(x^6)$$

To remove the `+ O(x^6)` representing series terms of higher order, we must convert the series to a "polynomial expression":

```
poly := convert( S, polynom );
```

$$poly := x - 1/6\ x^3 + 1/120\ x^5$$

But to define the Taylor polynomial as a function that is usable -- for instance, in plotting -- we must use a "dummy variable" trick:

```
P := (k,t) -> subs( x = t,
     convert( series(sin(x), x = 0, k+1), polynom ) );
```

to define the kth degree Taylor polynomial of the sine function. Then a command like

```
plot( {sin(x), P(3,x), P(5,x), P(7,x), P(9,x)},
              x = 0..3*Pi, y = -3..3 );
```

simultaneously plots the graph of the sine function and several of its Taylor polynomials.

65

For each function given in the exercises that follow, generate several pictures, each of which shows its graph and several Taylor polynomial approximations.

1. $f(x) = \sin x$

2. $f(x) = \cos x$

3. $f(x) = e^{-x}$

4. $f(x) = \dfrac{1}{1+x}$

5. $f(x) = \ln(1 + x)$

6. $f(x) = \dfrac{1}{1+x^2}$

7. $f(x) = \arctan x$

Project 33
Using *p*-Series to Approximate π

Reference: Section 11.5 of Edwards & Penney

You can use the *Maple* **sum** function as in Project 31 to calculate partial sums of the *p*-series

$$\zeta(p) = \sum_{n=1}^{\infty} \frac{1}{n^p} = 1 + \frac{1}{2^p} + \frac{1}{3^p} + \dots . \tag{1}$$

The notation $\zeta(p)$ -- read "zeta of p" -- is a standard abbreviation for the "zeta function" defined by the *p*-series in Eq. (1). If p is an even integer, then the exact value of $\zeta(p)$ is known as a rational multiple of π^p (whereas the exact value of $\zeta(p)$ is unknown for p an odd integer ≥ 3). In Problems 1 through 6 below use the listed known value of $\zeta(p)$ to see how accurately you can approximate the value

$$\pi = 3.14159\ 26535\ 89793\ \dots .$$

For instance, in Problem 2 you might calculate the kth partial sum s_k of $\sum 1/n^4$ for $k = 10, 20, 40, 80, \dots$ and observe how accurately $\sqrt[4]{90 s_k}$ approximates π.

1. $\zeta(2) = \dfrac{\pi^2}{6}$

2. $\zeta(4) = \dfrac{\pi^4}{90}$

3. $\zeta(6) = \dfrac{\pi^6}{945}$

4. $\zeta(8) = \dfrac{\pi^8}{9450}$

5. $\zeta(10) = \dfrac{\pi^{10}}{93555}$

6. $\zeta(12) = \dfrac{691\pi^{12}}{638512875}$

7.   Calculate partial sums of the harmonic series $\sum 1/n$ to approximate *Euler's constant* γ, which is defined by

$$\gamma = \lim_{n \to \infty} \left(1 + \frac{1}{2} + \frac{1}{3} + \ldots + \frac{1}{n} - \ln n \right) \approx 0.57722.$$

Unless you use a very powerful microcomputer, you likely will have to be content with only two or three decimal places of accuracy.

Project 34
Graphical Approximations
by Taylor Polynomials

Reference: Section 11.8 of Edwards & Penney

The heavier curve in Fig. 1 is the graph of the function

$$\frac{\sin x}{x} = 1 - \frac{x^2}{3!} + \frac{x^4}{5!} - \frac{x^6}{7!} + \ldots , \tag{1}$$

and the other graphs are those of its Taylor polynomial approximations of degrees $k = 2$, 4, 6, ... , 14, 16. The heavier curve in Fig. 2 is the graph of the antiderivative

$$f(x) = \int_0^x \frac{\sin t}{t} \, dt$$

$$= x - \frac{x^3}{3!3} + \frac{x^5}{5!5} - \frac{x^7}{7!7} + \ldots \tag{2}$$

of the function in Eq. (1); the other graphs are those of its Taylor polynomial approximations of degrees $k = 3, 5, 7, \ldots , 15, 17$.

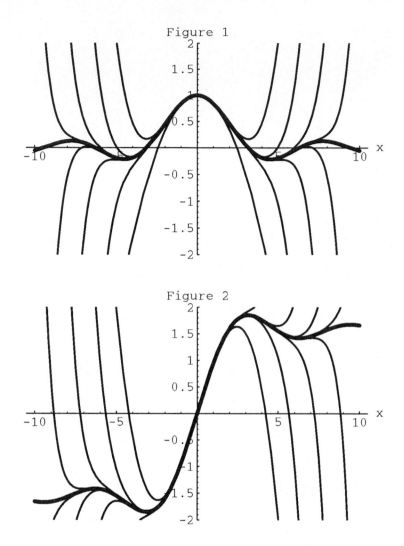

Figure 1

Figure 2

First verify the Taylor series expansions in Eqs. (1) and (2). Then use them to produce your own versions of Figs. 1 and 2. You may want to plot a smaller or larger number of Taylor polynomials, depending on the techniques you use. You may find useful the series summation techniques described in the previous projects (and especially the graphing techniques in Project 32).

Then do the same with each of the other function-antiderivative pairs indicated Problems 1-5 below.

1. $\quad f(x) = \int_0^x \sin t^3 \, dt$

2. $\quad f(x) = \int_0^x \exp(-t^3) \, dt$

68

3. $$f(x) = \int_0^x \frac{\arctan t}{t}\, dt$$

4. $$f(x) = \int_0^x \frac{1 - \exp(-t^2)}{t^2}\, dt$$

5. $$\tanh^{-1} x = \int_0^x \frac{dt}{1 - t^2}$$

Project 35
Using Power Series to
Evaluate Indeterminate forms

Reference: Section 11.9 of Edwards & Penney

This project explores the use of known power series such as

$$\sin x = x - \frac{x^3}{6} + \frac{x^5}{120} - \frac{x^7}{5040} + \dots,\tag{1}$$

$$\tan x = x + \frac{x^3}{3} + \frac{2x^5}{15} + \frac{17x^7}{315} + \dots,\tag{2}$$

$$\sin^{-1} x = x + \frac{x^3}{6} + \frac{3x^5}{40} + \frac{5x^7}{112} + \dots,\tag{3}$$

$$\tan^{-1} x = x - \frac{x^3}{3} + \frac{x^5}{5} - \frac{x^7}{7} + \dots\tag{4}$$

to evaluate certain indeterminate forms for which the use of l'Hôpital's rule would be inconvenient or impractical.

Problem 1

Evaluate $\displaystyle\lim_{x \to 0} \frac{\sin x - \tan x}{\sin^{-1} x - \tan^{-1} x}.$ (5)

When you substitute Eqs. (1) and (2) into the numerator and Eqs. (3) and (4) into the denominator, you will find that both the numerator series and the denominator series have leading terms that are multiples of x^3. Hence division of each term by x^3 leads quickly to the desired value of the indeterminate form. Can you explain why this implies that l'Hôpital's rule would have to be applied three times in succession? Would you like to calculate (by hand) the third derivative of the denominator in Eq. (5)?

69

Problem 2

Evaluate

$$\lim_{x \to 0} \frac{\sin(\tan x) - \tan(\sin x)}{\sin^{-1}(\tan^{-1} x) - \tan^{-1}(\sin^{-1} x)} \tag{6}$$

This is a much more substantial problem, and the use of a computer algebra system like *Maple* is virtually essential. First generate the Taylor series

$$\sin(\tan x) = x + \frac{x^3}{6} - \frac{x^5}{40} - \frac{55x^7}{1008} + \dots, \tag{7}$$

$$\tan(\sin x) = x + \frac{x^3}{6} - \frac{x^5}{40} - \frac{107x^7}{5040} + \dots, \tag{8}$$

$$\sin^{-1}(\tan^{-1} x) = x - \frac{x^3}{6} + \frac{13x^5}{120} - \frac{341x^7}{5040} + \dots, \tag{9}$$

$$\tan^{-1}(\sin^{-1} x) = x - \frac{x^3}{6} + \frac{13x^5}{120} - \frac{173x^7}{5040} + \dots. \tag{10}$$

Then show that when Eqs. (7) and (8) are substituted into the numerator in Eq. (6) and Eqs. (9) and (10) into the denominator, both resulting series have leading terms that are multiples of x^7. Hence division of each series by x^7 leads quickly to the desired evaluation. It follows that seven successive applications of l'Hôpital's rule would be required to evaluate the indeterminate form in Eq. (6). (Why?) Can you even conceive of calculating the seventh derivative of either the numerator or denominator in Eq. (6)? The seventh derivative of $\sin(\tan x)$ is a sum of sixteen terms, a typical one of which is $3696 \cos(\tan x) \sec^2 x \, \tan^2 x$.

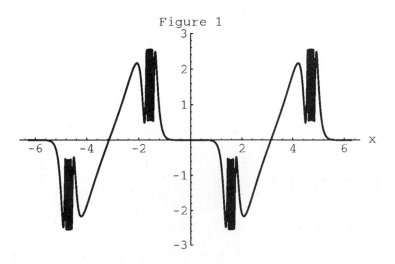

Figure 1

The fact that the power series for

$$f(x) = \sin(\tan x) - \tan(\sin x) \tag{11}$$

70

and

$$g(x) = \sin^{-1}(\tan^{-1} x) - \tan^{-1}(\sin^{-1} x) \qquad (12)$$

both begin with the term involving x^7 means, geometrically, that the graphs of both are exceedingly "flat" near the origin (Figs. 1 and 2).

The graph of $y = f(x)$ in Fig. 1 is especially exotic. Can you explain the conspicuous oscillations that appear in certain sections of the graph?

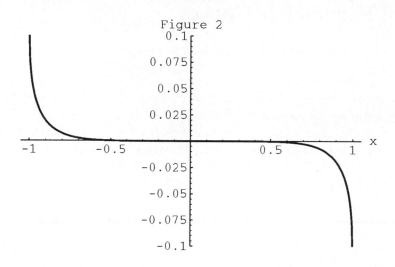

Figure 2

Chapter 12

Parametric Curves and Vectors in the Plane

Project 36
Graphs of Parametric Curves

Reference: Section 12.1 of Edwards & Penney

The most fun in plotting parametric curves comes from trying your own hand --
especially if you have a graphics calculator or computer with graphing utility to do the
real work. Try various values of the constants a, b, p, ... in the examples below.
When sines and cosines are involved, the interval $0 \le t \le 2\pi$ is a reasonable one to try
first. You will need to experiment with different windows to find one that shows the
whole curve (or the most interesting part of it). In each project, first define the
appropriate coordinate functions $x(t)$ and $y(t)$, and then use a parametric plot command
of the form

```
plot( [ x(t), y(t), t = a..b ] );
```

Project 36A
Given: $x = at - b \sin t$, $y = a - b \cos t$. This *trochoid* is traced by a point P on a
solid wheel of radius a as it rolls along the x-axis; the distance of P from the center of
the wheel is $b > 0$. (The graph is a cycloid if $a = b$.) Try both cases $a > b$ and $a < b$.

Project 36B
Given:

$$x = (a - b)\cos t + b\cos\left(\frac{a-b}{b}t\right),$$

$$y = (a - b)\sin t - b\sin\left(\frac{a-b}{b}t\right).$$

This is a *hypocycloid* -- the path of a point P on a circle of radius b that rolls along the
inside of a circle of radius $a > b$. Figure 1 shows the case $b = a/5$.

Project 36C
Given:

$$x = (a + b)\cos t - b\cos\left(\frac{a+b}{b}t\right),$$

$$y = (a + b)\sin t - b\sin\left(\frac{a+b}{b}t\right).$$

This is an *epicycloid* traced by a point P on a circle of radius b that rolls along the
outside of a circle of radius $a > b$. Figure 2 shows the case $b = a/5$.

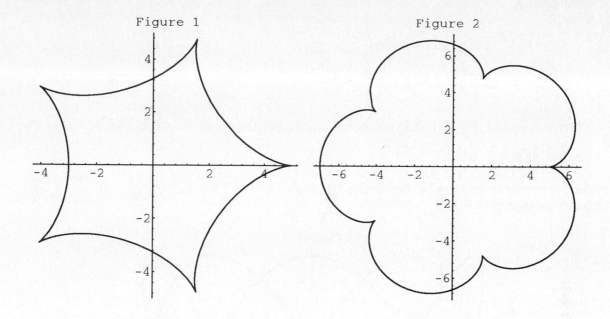

Figure 1

Figure 2

Project 36D

Given:

$$x = a \cos t - b \cos \frac{at}{2},$$

$$x = a \sin t - b \sin \frac{at}{2}.$$

This is an *epitrochoid* -- it is to an epicycloid what a trochoid is to a cycloid. With $a = 8$ and $b = 5$ you should get the curve shown in Fig. 3.

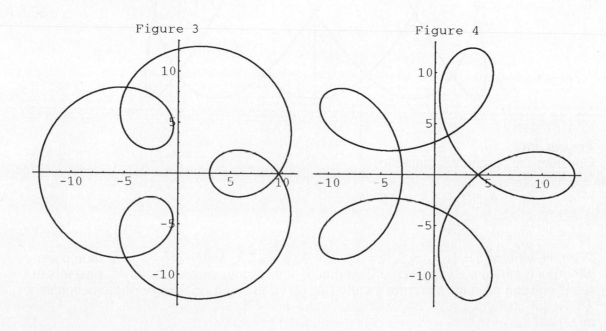

Figure 3

Figure 4

Project 36E
Given:

$$x = a \cos t + b \cos \frac{at}{2},$$

$$x = a \sin t - b \sin \frac{at}{2}.$$

With $a = 8$ and $b = 5$ this *hypotrochoid* looks like the curve shown in Fig. 4.

Project 36F
Given: $x = \cos at$, $y = \sin bt$. These are the *Lissajous curves* that typically appear on oscilloscopes in physics or electronics laboratories. The Lissajous curve with $a = 3$ and $b = 5$ is shown in Fig. 5.

Figure 5

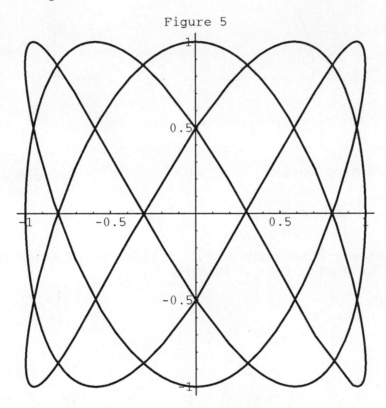

Project 36G
Consider the parametric equations

$$x = a \cos t - b \cos pt,$$

$$y = c \sin t - d \sin qt.$$

With the values $a = 16$, $b = 5$, $c = 12$, $d = 3$, $p = 47/3$, and $q = 44/3$ you should see why this is called a *slinky curve*. Experiment with various combinations of constants to see if you can produce a prettier picture (the criterion being "in the eye of the beholder").

Numerical Approximation
of Parametric Arc Length

Reference: Section 12.2 of Edwards & Penney

These projects call for the use of numerical integration techniques (as described in Projects 14 and 16) to approximate the parametric arc-length integral

$$s = \int_a^b [x'(t)^2 + y'(t)^2]^{1/2} \, dt. \tag{1}$$

Consider the ellipse with equation

$$\frac{x^2}{a^2} + \frac{y^2}{b^2} = 1 \qquad (a > b) \tag{2}$$

and eccentricity $\varepsilon = \sqrt{1 - (a/b)^2}$. Substitute the parametrization

$$x = a \cos t, \quad y = b \sin t \tag{3}$$

into Eq. (1) to show that the perimeter of the ellipse is given by the *elliptic integral*

$$p = 4a \int_0^{\pi/2} \sqrt{1 - \varepsilon^2 \cos^2 t} \, dt. \tag{4}$$

This integral is known to be nonelementary if $0 < \varepsilon < 1$. A common simple approximation to it is

$$p \approx \pi(A + R), \tag{5}$$

where

$$A = \frac{a+b}{2} \qquad \text{and} \qquad R = \sqrt{\frac{a^2 + b^2}{2}}$$

denote the *arithmetic mean* and *root-square mean*, respectively, of the semiaxes a and b of the ellipse.

Project 37A

Consider the ellipse whose major and minor semiaxes a and b are, respectively, the largest and smallest nonzero digits of your student I.D. number. For this ellipse, estimate the perimeter given in Eq. (4) by using the right- and left-endpoint approximations R_n and L_n, the trapezoidal approximation $T_n = (R_n + L_n)/2$, the midpoint approximation M_n, and Simpson's approximation S_n. In each case compare these approximations using $n = 10, 20, 40, \ldots$ subintervals. Construct a table showing your results along with the simple estimate in Eq. (5).

Project 37B

If we ignore the perturbing effects of the sun and the planets other than the earth, the orbit of the moon is an almost perfect ellipse with the earth at one focus. Assume that this ellipse has semimajor axis $a = 384,403$ km (exactly) and eccentricity $\varepsilon = 0.0549$ (exactly). Approximate the length p of this ellipse (Eq. (4)) to the nearest meter.

Project 37C

Suppose you are designing an elliptical auto racetrack. Pick semiaxes for *your* racetrack so that its perimeter will be somewhere between a half mile and two miles. Your task is to construct a table with *time* and *speed* columns that an observer can use to determine the average speed of a particular car as it circles the track. The times listed in the first column should correspond to speeds up to perhaps 150 mi/h. The observer clocks a car's circuit of the track and locates its time for that lap in the first column of the table. The corresponding figure in the second column then gives the car's average speed (in mi/h) for that circuit of the track. Your report should include a nice convenient table to use this way -- so you can successfully sell it to racetrack patrons attending the auto races.

Chapter 13

Vectors, Curves, and Surfaces in Space

Project 38
Investigating the Curve of a Baseball Pitch

Reference: Section 13.4 of Edwards & Penney

Have you ever wondered whether a curveball really curves, or whether it's some sort of optical illusion? In this project you'll use calculus to settle the matter.

Suppose that a pitcher throws a ball toward home plate (60 ft away) and gives it a spin of S revolutions per second counter-clockwise (as viewed from above) about a vertical axis through the center of the ball. This spin is described by the *spin vector* $\mathbf{S}$ that points along the axis of revolution in the right-handed direction and has length S.

We know from studies of aerodynamics that this spin causes a difference in air pressure on the sides of the ball toward and away from this spin. This pressure difference results in a *spin acceleration*

$$\mathbf{a}_S \;=\; c\mathbf{S}\times\mathbf{v} \qquad \text{(vector product)} \tag{1}$$

of the ball (where c is an empirical constant). The total acceleration of the ball is then

$$\mathbf{a} \;=\; c\mathbf{S}\times\mathbf{v} - g\mathbf{k}, \tag{2}$$

where $g \approx 32$ ft/s^2 is the gravitational acceleration. Here we will ignore any other effects of air resistance.

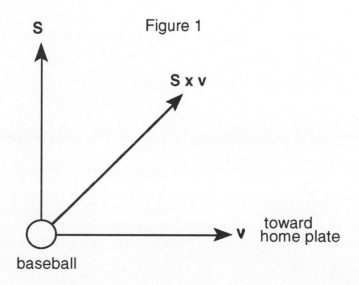

Figure 1

With the spin vector $\mathbf{S} = S\mathbf{k}$ pointing upward as in Fig. 1, show first that

$$\mathbf{S} \times \mathbf{v} = -Sv_y \mathbf{i} + Sv_x \mathbf{j}, \tag{3}$$

where v_x is the component of $\mathbf{v}$ in the x-direction, v_y is the component of $\mathbf{v}$ in the y-direction.

For a ball pitched along the x-axis, v_x is much larger than v_y, so we may assume that $v_y \approx 0$. Hence the approximation

$$\mathbf{S} \times \mathbf{v} \approx Sv_x \mathbf{j} \tag{4}$$

is sufficiently accurate for our purposes. We may then take the acceleration vector of the ball to be

$$\mathbf{a} = cSv_x \mathbf{j} - g\mathbf{k}. \tag{5}$$

Now suppose that the pitcher throws the ball from the initial position $x_0 = y_0 = 0$, $z_0 = 5$ (ft), with initial velocity vector

$$\mathbf{v}_0 = 120\mathbf{i} - 3\mathbf{j} + 4\mathbf{k} \tag{6}$$

(with components in feet per second, so $v_0 \approx 120$ ft/s, about 82 mi/h) and with a spin of $S = 40$ rev/s. A reasonable value of c is

$$c = 0.005 \text{ ft/s}^2 \text{ per ft/s of velocity and revolutions/second of spin,}$$

although the precise value depends on whether the pitcher has (accidentally, of course) scuffed the ball or administered to it some foreign substance.

Show first that these values of the parameters yield

$$\mathbf{a} = 24\mathbf{j} - 32\mathbf{k} \tag{7}$$

for the ball's acceleration vector. Then integrate twice in succession to find the ball's position vector

$$\mathbf{r}(t) = x(t)\mathbf{i} + y(t)\mathbf{j} + z(t)\mathbf{k}. \tag{8}$$

Use your results to fill in the following table, giving the pitched ball's horizontal deflection y and height z (above the ground) at one-eighth second intervals.

t (s)	x (ft)	y (ft)	z (ft)
0.0	0	0	5
0.125	15	?	?
0.25	30	?	?
0.375	45	?	?
0.50	60	?	?

Suppose that the batter gets a "fix" on the pitch by observing the ball during the first quarter-second and prepares to swing. Does the pitch still appear after 0.25 s to be straight on target toward home plate at a height of 5 ft?

What happens to the ball during the final quarter-second of its approach to home plate -- *after* the batter has begun to swing the bat? What are the ball's horizontal and vertical deflections during this brief period? What is *your* conclusion? Does the pitched ball really "curve" or not?

Using *Maple*

The simplifying assumption in (4) makes the acceleration vector constant, so it is possible to start with Eq. (7) and obtain the coordinate functions $x(t)$, $y(t)$, $z(t)$ in Eq. (8) after two successive routine integrations (by hand). Now we discuss the use of *Maple* to solve the baseball curve problem without this assumption. With the constant values

```
c := 0.005;     S := 40;     g := 32;
```

show first that (2) and (3) yield the acceleration equations

```
xDE :=      D(D(x))(t) = -c*S*D(y)(t);
yDE :=      D(D(y))(t) =  c*S*D(x)(t);
zDE :=      D(D(z))(t) = -g;
```

We have also the initial position coordinates

$$x(0) = 0, \quad y(0) = 0, \quad z(0) = 5,$$

and the initial velocity components

$$x'(0) = 120, \quad y'(0) = -3, \quad z'(0) = 4.$$

Now we use *Maple's* **dsolve** function to find the baseball's coordinate functions exactly. The first two differential equations do not involve $z(t)$, so the ball's x- and y-coordinates are given by

```
xyCoords := dsolve( {xDE, yDE,
                    x(0) = 0, y(0) = 0,
                    D(x)(0) = 120, D(y)(0) = -3},
                    {x(t),y(t)} );
```

The third differential equation involves only $z(t)$, so the ball's z-coordinate is given by

```
zCoord := dsolve( {zDE, z(0) = 5, D(z)(0) = 4}, z(t) );
```

Upon executing these commands, you will find that $x(t)$ and $y(t)$ are simple combinations of $\sin 0.2t$ and $\cos 0.2t$ while $z(t)$ is a quadratic polynomial in t. Then the ball's position after 0.25 seconds is given by

```
evalf(subs( t = 0.25, xyCoords ));
evalf(subs( t = 0.25,   zCoord ));
```

```
              {x(.25) = 30.00624765, y(.25) = .0001563 }
                        z(.25) = 5.0000
```

Thus the ball appears headed straight toward home plate. But after 0.5 seconds, its position is

```
evalf(subs( t = 0.5, xyCoords ));
evalf(subs( t = 0.5,   zCoord  ));
```

$$\{\ x(.5) = 59.97498751,\ y(.5) = 1.4999996,\}$$
$$z(.5) = 3.00$$

so as it approaches home plate, it appears to have "broken" 2 ft downward and 1.5 ft outside home plate.

Project 39
Graphical Construction of Lathe Objects

Reference: Section 13.7 of Edwards & Penney

A surface of revolution around the *z*-axis is readily described in cylindrical coordinates. The surface obtained by revolving the curve $r = f(z)$, $a \le z \le b$ is given by

$$x = f(z) \cos \theta, \quad y = f(z) \sin \theta, \quad z = z$$

for $0 \le \theta \le 2\pi$.

Figure 1

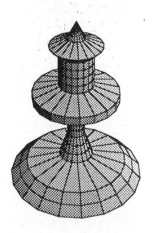
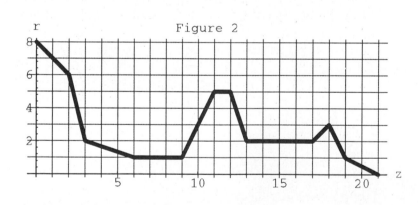

The figure of the chess piece shown in Fig. 1 was plotted by revolving the polygonal curve of Fig. 2 around the (vertical) *z*-axis in space. The first step in the construction was to draw the desired curve $r = f(z)$ on a piece of graph paper, then measure the coordinates $(0, 8)$, $(2, 6)$, $(3, 2)$, ..., $(18, 3)$, $(19, 1)$, and $(21, 0)$ of the vertices. The point-slope formula was used to define $r = f(z)$ on successive intervals corresponding to straight line segments of the polygonal curve:

80

$$r = 8 - z \qquad \text{on } [0, 2],$$
$$r = 14 - 4z \qquad \text{on } [2, 3],$$

.
.
.

$$r = 5 \qquad\qquad \text{on } [11, 12],$$

.
.
.

$$r = 39 - 2z \qquad \text{on } [18, 19],$$
$$r = (21 - z)/2 \qquad \text{on } [19, 21].$$

A "construction curve" like the one in Fig. 2 can be used as input to an automatic lathe, which will then follow it as a pattern to cut the corresponding solid of revolution from a cylinder of wood or metal. Some years ago, many freshman engineering students had to produce a metal lamp stem in this fashion -- from the original graph paper design to the actual operation of a manual lathe.

Your assignment in this project is simpler: Merely produce a *figure* showing such a "lathe object." Think of a lamp stem, a table leg, or perhaps a more complicated chess piece than the pawn shown in Fig. 1. Once the construction curve $r = f(z)$ has been defined, the figure can be generated by the *Maple* parametric plot command

```
plot3d( [ f(z)*cos(t), f(z)*sin(t), z ],
                    z = a..b, t = 0..2*Pi );
```

If a computer system that plots parametric surfaces were unavailable, you still could draw a lathe object corresponding to a *polygonal* construction curve. You would begin by drawing horizontal circles corresponding to the vertices (as ellipses in perspective).

Chapter 14

Partial Differentiation

Project 40
Computer Plotting of
Three-Dimensional Surfaces

Reference: Section 14.2 of Edwards & Penney

Plotting surfaces with a computer graphing program can help you develop a "feel" for graphs of functions of two variables. To use *Maple* to plot the surface $z = f(x, y)$ over the base rectangle $a \le x \le b$, $c \le y \le d$, you need only first define the function

```
f := (x,y) -> x^2 - y^2;              #  for instance
```

and then enter the three-dimensional plot command

```
plot3d( f(x,y), x = a..b, y = c..d );
```

With $a = c = -1$ and $b = d = 1$ (and the options **axes = BOXED, style = PATCH**) the result looks as shown in Fig. 1.

Figure 1

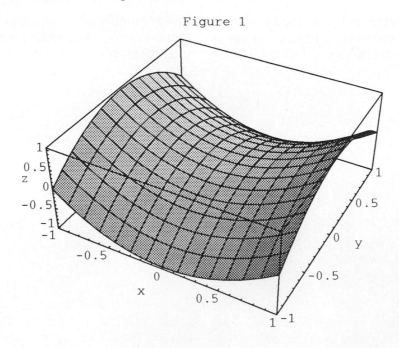

   To begin, graph some of the functions listed next over rectangles of various sizes to see how the scale affects the picture. Similarly, vary the numerical parameters p, q, and r and note the resulting changes in the graph. Then make up some functions of your own for experimentation. If you have a computer connected to a printer, assemble a portfolio of your most interesting examples.

82

$$f(x, y) = p \cos qx$$
$$f(x, y) = p \cos qy$$
$$f(x, y) = \sin px \sin qy$$
$$f(x, y) = p + qx^2 \qquad \text{(Use negative } and \text{ positive}$$
$$f(x, y) = p + qy^2 \qquad\qquad \text{values of } p \text{ and } q \text{ in}$$
$$f(x, y) = px^2 + qy^2 \qquad\qquad \text{these three examples.)}$$
$$f(x, y) = px^2 + qxy + ry^2$$
$$f(x, y) = \exp(px^2 - qy^2)$$
$$f(x, y) = (px^2 + qxy + ry^2)\exp(-x^2 - y^2)$$

Project 41
Graphical Investigation of Extreme Values on a Disk

Reference: Section 14.5 of Edwards & Penney

Let R denote the unit circular disk $x^2 + y^2 \leq 1$ bounded by the unit circle C in the xy-plane. Suppose that we want to find the minimum and maximum values of the function

$$z = f(x, y) = 3x^2 + 4xy - 5y^2 \tag{1}$$

at points of R.

We look first for interior possibilities. The equations

$$\frac{\partial f}{\partial x} = 6x + 4y = 0,$$

$$\frac{\partial f}{\partial y} = 4x - 10y = 0$$

have only the trivial solution $x = y = 0$, so the only interior possibility is $f(0, 0) = 0$.

To investigate the boundary possibilities we use the polar-coordinates parametrization

$$x = \cos t, \quad y = \sin t \qquad (0 \leq t \leq 2\pi) \tag{2}$$

of the boundary circle C. Substitution of (2) into Eq. (1) then yields the single-variable

$$z = g(t) = 3 \cos^2 t + 4 \cos t \sin t - 5 \sin^2 t, \tag{3}$$

whose extreme values we now seek.

The graph $z = g(t)$ plotted in Fig. 1 reveals a positive maximum value and a negative minimum value, each occurring at two different points in $[0, 2\pi]$. The zooms indicated in Figs. 2 and 3 yield (accurate to two decimal places) the approximate maximum value $g(0.23) \approx 3.47$ and the approximate minimum value $g(1.80) \approx -5.47$ attained by $f(x, y)$ at points of the disk R. You can check these graphical solutions using the *Maple* commands

```
root := fsolve( D(g)(t), t, t = 0..0.5 );
g(root);
```

83

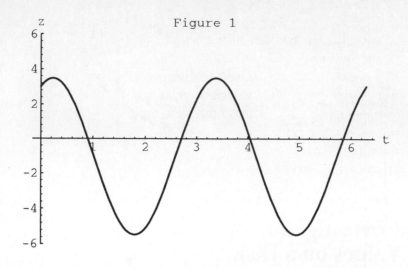

Figure 1

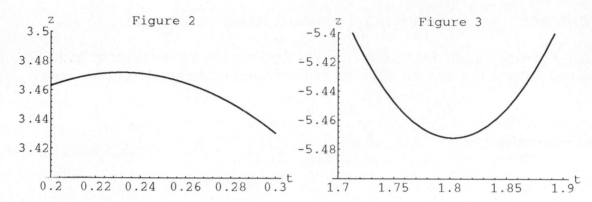

Figure 2   Figure 3

In Problems 1 through 3, find similarly the approximate maximum and minimum values attained by the indicated function $f(x, y)$ at points of the unit disk R: $x^2 + y^2 \leq 1$. Let p, q, and r denote three selected integers, such as the last three nonzero digits of your student I.D. number.

1. $f(x, y) = px + qy + r$

2. $f(x, y) = px^2 + qxy + ry^2$

3. $f(x, y) = px^4 + qy^4 - rx^2y^2$

Project 42
Computer Solution of Lagrange Multiplier Problems

Reference: Section 14.9 of Edwards & Penney

Project 42A
Figure 1 shows an alligator-filled moat of width $w = 10$ ft bounded on each side by a wall of height $h = 6$ ft. Soldiers plan to bridge this moat by scaling a ladder that is placed across the wall as indicated and anchored at the ground by a handy boulder, with

the upper end directly above the wall on the opposite side of the moat. What is the minimal length L of a ladder that will suffice for this purpose? We outline two approaches.

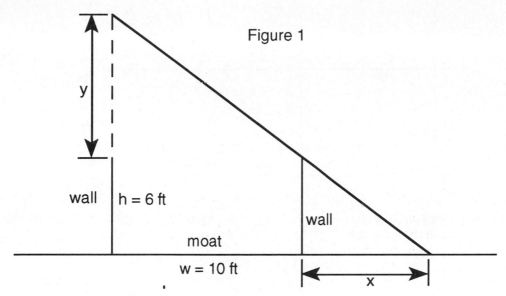

Figure 1

With a Single Constraint

Apply the Pythagorean theorem and the proportionality theorem for similar triangles to show that you need to minimize the (ladder-length squared) function

$$f(x, y) = (x + 10)^2 + (y + 6)^2$$

subject to the constraint

$$g(x, y) = xy - 60 = 0.$$

Then apply the Lagrange multiplier method to derive the fourth-degree equation

$$x^4 + 10x^3 - 360x - 3600 = 0. \tag{1}$$

Using *Maple* you can first define the functions

```
f   :=   (x,y) -> (x + 10)^2 + (y + 6)^2;
g   :=   (x,y) -> x*y - 60;
```

Then set up the Lagrange multiplier equations:

```
eq1 :=    diff( f(x,y), x) = lambda*diff( g(x,y), x);
eq2 :=    diff( f(x,y), y) = lambda*diff( g(x,y), y);
```

And finally eliminate `y` and `lambda` by **subs**tituting `y = 60/x` into the equation

```
solve( eq1, lambda) = solve( eq2, lambda);
```

You can approximate the pertinent (positive) solution of Eq. (1) graphically (Fig. 2). You may even be able to solve this equation manually -- if you can first spot an integer solution (which must be an integral factor of the constant term 3600).

85

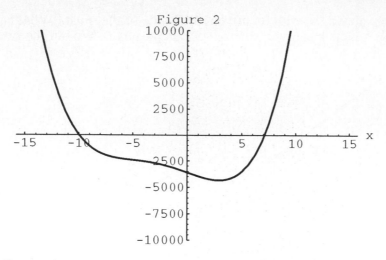

Figure 2

With Two Constraints

Here is an alternative approach to the moat problem. With $z = L$ for the length of the ladder, observe directly from Fig. 1 that you need to minimize the function

$$f(x, y, z) = z$$

subject to the two constraints

$$g(x, y, z) = xy - 60 = 0,$$

$$h(x, y, z) = (x + 10)^2 + (y + 6)^2 - z^2 = 0.$$

This leads to a system of five equations in five unknowns $(x, y, z,$ and the two Lagrange multipliers). Perhaps you can then use the *Maple* command

```
fsolve( {equations}, {unknowns} );
```

to solve a list of equations for its list of unknowns.

For your own personal moat problem, you might choose w and $h < w$ as the two largest distinct digits in your student I.D. number.

Project B

Let u and x be numbers with $0 < u < x < 1$, and define

$$y = \sqrt{1 - x^2}, \qquad v = \sqrt{1 - u^2}.$$

Then consider the 14-sided polygon that has has vertices $(0, \pm 1)$, $(\pm x, \pm y)$, $(\pm u, \pm v)$, and $(\pm u, \pm y)$ and thus is almost inscribed in the unit circle. When this polygon is revolved around the y-axis, it generates the "spindle" solid illustrated in Fig. 3, which consists of a solid cylinder of radius x, two solid cylinders of radius u, and two solid cones. The problem is to determine x, y u, and v in order to *maximize* the volume of this spindle.
First express the volume V of the spindle as a function

$$V = f(x, y, u, v)$$

86

Figure 3

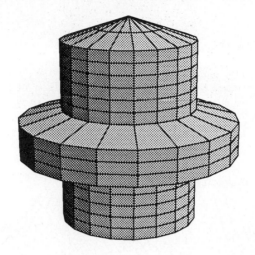

of four variables. The problem then is to maximize $f(x, y, u, v)$ subject to the two constraints

$$g(x, y, u, v) = x^2 + y^2 - 1 = 0,$$

$$h(x, y, u, v) = u^2 + v^2 - 1 = 0.$$

The corresponding Lagrange multiplier condition takes the form

$$\nabla f = \lambda \, \nabla g + \mu \, \nabla h$$

where $\nabla f = \langle f_x, f_y, f_u, f_v \rangle$ and ∇g and ∇h are similar four-vectors of partial derivatives.

All this results in a system of six equations in the six unknowns x, y, u, v, λ, μ. You can set up this system manually, but you will surely need *Maple* to solve it.

Project 43
Numerical Classification of Critical Points

Reference: Section 14.10 of Edwards & Penney

Project 43A
Consider the function f of two variables defined by

$$f(x,y) = 10\left(x^3 + y^5 + \tfrac{1}{5}x\right)\exp(-x^2 - y^2) + \tfrac{1}{3}\exp[-(x-1)^2 - y^2], \tag{1}$$

whose graph is shown in Fig. 1. Because $f(x,y) \to 0$ as $x, y \to \pm\infty$, the surface $z = f(x, y)$ must have a highest point and a lowest point. Your task is to find them.

Figure 1

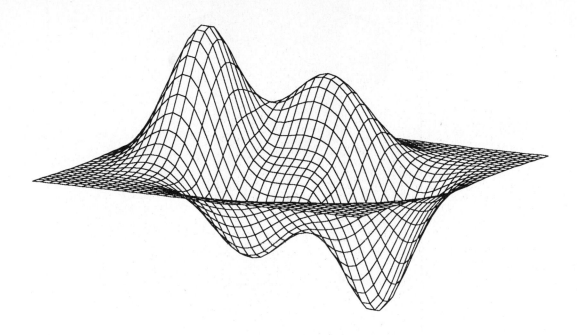

Six critical points are visible: two local maxima, two saddle points, and two local minima. The problem is to calculate the locations of these critical points. Show first that when you calculate the two partial derivatives f_x and f_y, equate both to zero, and then remove the common factor $\exp(-x^2-y^2)$ from each resulting equation, you get the two equations

$$-\tfrac{2}{3}(x-1)e^{2x-1} - 20x^4 + 26x^2 - 20xy^5 + 2 = 0, \tag{2}$$

$$-\tfrac{2}{3}ye^{2x-1} - 20x^3y - 4xy - 20y^6 + 50y^4 = 0 \tag{3}$$

to be solved for the coordinates x and y of the critical points of the function f.

Although these equations look intimidating, they have two redeeming features that permit you to solve them by essentially single-variable techniques:

- Note that $y = 0$ satisfies Eq. (3). So you can find the critical points that lie on the x-axis by substituting $y = 0$ into Eq. (2), then solving the remaining equation in x by graph-and-zoom techniques.

88

You can solve Eq. (2) for y in terms of x. When you substitute the result into Eq. (3), you get another equation in x that can be solved by graph-and-zoom techniques. This gives the remaining critical points.

Figure 2

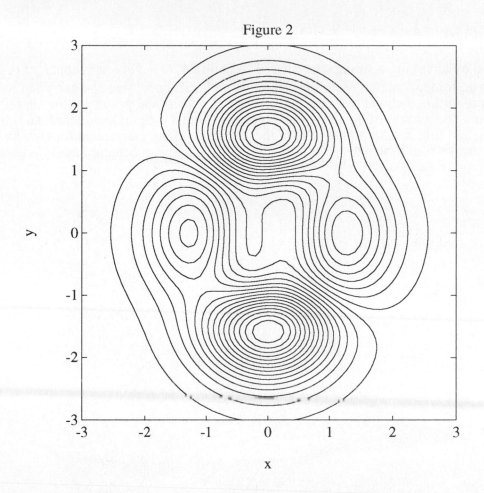

Here is an alternative, two-variable approach. You might use the (*Maple* V Release 2) command

```
plot3d( f(x,y), x =-3..3, y =-3..3, style = CONTOUR );
```

to generate a contour plot of $f(x, y)$ resembling the one shown in Fig. 2. There it is apparent that the function f has local max-min points with the (very) approximate locations $(\pm 1.25, 0)$ and $(0, \pm 1.5)$ as well as saddle points with the approximate locations $(\pm 1, \pm 1)$. Having located each of these critical points approximately, you can then solve for each more precisely using a command of the form

```
fsolve( {diff(f(x,y),x) = 0, diff(f(x,y),y) = 0},
            {x,y}, {x=x1..x2, y=y1..y2} );
```

which "sets the two partial derivatives equal to 0" and solves for a critical point (x, y) in the indicated intervals.

Project 43B
Figure 3 shows the graph of the new function $g(x, y)$ defined by the formula

$$g(x,y) = 10\left(x^3 + y^5 - \tfrac{1}{5}x\right)\exp(-x^2 - y^2) + \tfrac{1}{3}\exp[-(x-1)^2 - y^2] \tag{4}$$

obtained by changing a single sign (that of $\tfrac{1}{5}x$) in Eq. (1). Now it is apparent that there is some additional "action" near the origin. With persistence, you can carry out the procedure outlined above to locate and analyze all the critical points of this altered function. You should find that g has ten critical points -- three local maxima, three local minima, and four saddle points. Six of these critical points are located roughly as in Project A; the four new ones are located within a unit square centered at the origin.

Figure 3

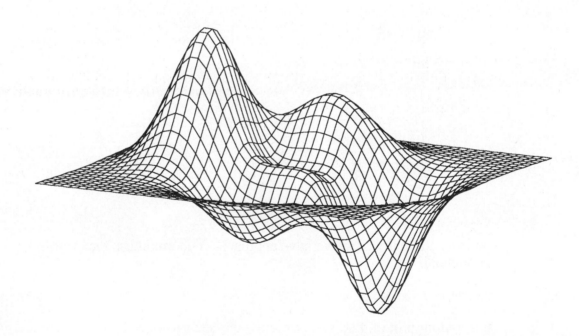

REMARK We first saw the function $g(x, y)$ in a brochure describing the *MATLAB* software system for interactive numerical computation (which was used to generate Figs. 1-3 above). The discovery of the function $f(x, y)$, whose critical point structure is somewhat simpler, was the serendipitous result of a typographical error made the first time we attempted to investigate the function $g(x, y)$.

Chapter 15

Multiple Integrals

Project 44
Numerical Approximation of Double Integrals

Reference: Section 15.1 of Edwards & Penney

This project explores the *midpoint approximation* to the double integral

$$I = \iint_R f(x,y)\, dA \tag{1}$$

of the function $f(x, y)$ over the plane rectangle $R = [a, b] \times [c, d]$. To define the midpoint approximation, let $[a, b]$ be subdivided into m subintervals all having the same length $h = \Delta x = (b - a)/m$, and let $[c, d]$ be subdivided into n subintervals all having the same length $k = \Delta y = (d - c)/n$. For each i and j ($1 \leq i \leq m$ and $1 \leq j \leq n$), let u_i and v_j denote the *midpoints* of the ith subinterval $[x_{i-1}, x_i]$ and the jth subinterval $[y_{j-1}, y_j]$, respectively. Then the corresponding **midpoint approximation** to the double integral I is the sum

$$S_{mn} = \sum_{i=1}^{m} \sum_{j=1}^{n} f(u_i, v_j)\, hk. \tag{2}$$

Example
If $m = 3$ and $n = 2$, then

$$h = \Delta x = \frac{b-a}{3}, \qquad k = \Delta y = \frac{d-c}{2},$$

and

$$S_{32} = hk\,[f(u_1, v_1) + f(u_2, v_1) + f(u_3, v_1)$$
$$+ f(u_1, v_2) + f(u_2, v_2) + f(u_3, v_2)].$$

The midpoint approximation to the integral in Eq. (1) is readily calculated using *Maple*. Suppose the function

```
f := (x,y) -> x^2 * y^3;        #  for instance
```

and the limits

```
a := 1;    b := 2;
c := 3;    d := 4;
```

have been entered. Then the commands

```
m := 5;   n := 5;          #  numbers of subintervals

h := (b - a)/m;
k := (d - c)/n;

midpointSum :=
    sum( sum( f(a - h/2 + i*h,c - k/2 + j*k),
                       i = 1..m ), j = 1..n ) *h*k ;
```

yield the desired midpoint approximation.

For each of the double integrals in Problems 1 through 6, first calculate the midpoint approximation S_{mn} with the indicated values of m and n. Then try larger values. Compare each numerical approximation with the exact value of the integral.

1. $\int_0^1 \int_0^1 (x+y) \, dy \, dx$, $m = n = 2$

2. $\int_0^3 \int_0^2 (2x+3y) \, dy \, dx$, $m = 3$, $n = 2$

3. $\int_0^2 \int_0^2 xy \, dy \, dx$, $m = n = 2$

4. $\int_0^1 \int_0^1 x^2 y \, dy \, dx$, $m = n = 3$

5. $\int_0^{\pi/2} \int_0^{\pi/2} \sin x \sin y \, dy \, dx$, $m = n = 2$

6. $\int_0^{\pi/2} \int_0^1 \frac{\cos x}{1+y^2} \, dy \, dx$, $m = n = 2$

Project 45
Design of Optimal
Downhill Race-Car Wheels

Reference: Section 15.5 of Edwards & Penney

To see moments of inertia in action, suppose that your club is designing an unpowered race car for the annual downhill derby. You have a choice of solid wheels, bicycle wheels with thin spokes, or even solid spherical wheels (like giant ball bearings). The question is this: Which wheels will make the race car go fastest?

Imagine an experiment in which you roll various sorts of wheels down an inclined plane to see which reaches the bottom the fastest. Suppose that a wheel of radius a and mass M starts from rest at the top (at height h) with potential energy $PE = Mgh$ and reaches the bottom with angular speed ω and (linear) velocity $v = a\omega$. Then (by conservation of energy) the wheel's initial potential energy has been transformed to a sum

$$Mgh = KE_{tr} + KE_{rot}$$

of translational kinetic energy

$$KE_{tr} = \tfrac{1}{2} M v^2$$

and rotational kinetic energy

$$KE_{rot} = \tfrac{1}{2} I_0 \omega^2 = \frac{I_0 v^2}{2a^2} \tag{1}$$

(a consequence of Eq. (9) in the reference for this section). Thus

$$Mgh = \tfrac{1}{2} M v^2 + \frac{I_0 v^2}{2a^2}. \tag{2}$$

The following problems explore the implications of this formula.

1. Suppose that the wheel's (polar) moment of inertia is given by

$$I_0 = kMa^2 \tag{3}$$

for some constant k. (For instance, $k = \tfrac{1}{2}$ for a wheel in the shape of a uniform solid disk -- by Example 10 in the reference for this section.) Then deduce from Eq. (2) that

$$v = \sqrt{\frac{2gh}{1+k}}. \tag{4}$$

Thus the smaller is k (and hence the smaller the wheel's moment of inertia), the faster the wheel rolls down the incline.

In Problems 2 through 8 take $g = 32$ ft/s², and assume that the vertical drop of the incline is $h = 100$ ft.

2. Why does it follow from Eq. (4) that, whatever the design, the maximum velocity a circular wheel can attain on this incline is 80 ft/s (just under 55 mi/h)?

3. If the wheel is a uniform solid disk (like an old-fashioned wagon wheel) with $I_0 = \tfrac{1}{2} Ma^2$ (by Eq. (3) with $k = \tfrac{1}{2}$), what is its speed v at the bottom of the incline?

4. Answer Problem 3 if, instead, the wheel is shaped like a narrow bicycle tire with its entire mass effectively concentrated at the distance a from its center. In this case, $I_0 = Ma^2$. (Why?)

5. Answer Problem 3 if, instead, the wheel is shaped like an annular ring (or washer) with outer radius a and inner radius b.

Before attempting Problems 6 through 8, wait until you have studied triple integration in spherical coordinates. In Problem 6, for instance, you need to know that the moment of inertia about a vertical axis of a uniform solid sphere of mass M and radius a is given by

$$I_z = \frac{2}{5} Ma^2.$$

It would be informative for you to apply *Maple*'s symbolic integration capabilities to derive this result. In each of these next three problems, the question is the same: What is the velocity of the wheel when it reaches the bottom of the incline?

6.   The wheel is a uniform solid sphere of radius a (a "giant ball bearing").

7.   The wheel is a very thin spherical shell whose entire mass is effectively concentrated at the distance a from its center.

8.   The wheel is a spherical shell with outer radius a and inner radius $b = a/2$.

    Finally, what is your conclusion? What is the shape of the wheels that yield the fastest downhill race car?

Project 46
Moment of Inertia and the Interior of the Earth

Reference: Section 15.7 of Edwards & Penney

If the earth were a perfect sphere with radius $R = 6370$ km, *uniform* density δ, and mass $M = \frac{4}{3} \delta \pi R^3$, then (according to the result cited in Project 45) its moment of inertia about its polar axis would be $I = \frac{2}{5} MR^2$. In actuality, however, it turns out that

$$I = kMR^2 \tag{1}$$

with $k < 0.4 = \frac{2}{5}$. The reason is that, instead of having a uniform interior, the earth has a dense core -- a solid sphere -- covered with a lighter mantle -- a spherical shell -- a few thousand kilometers thick. The density of the core is

$$\delta_1 \approx 11 \times 10^3 \quad \text{kg/m}^3$$

and that of the mantle is

$$\delta_2 \approx 5 \times 10^3 \quad \text{kg/m}^3.$$

    The numerical value of k in Eq. (1) can be determined from certain earth satellite observations. If the earth's polar moment of inertia I and mass M (for the core-mantle model) are expressed in terms of the (unknown) radius x of the spherical core, then substitution of these expressions in to Eq. (1) yields an equation that can be solved for x.
    Show that this equation can be written in the form

$$2(\delta_1 - \delta_2)x^5 - 5k(\delta_1 - \delta_2)R^2x^3 + (2 - 5k)\delta_2 R^5 = 0. \tag{2}$$

Assuming that $k = 0.371$, solve this equation (graphically or numerically) to find x, and thereby determine the thickness of the earth's mantle.

Project 47
Computer Plotting of Parametric Surfaces

Reference: Section 15.8 of Edwards & Penney

Parametric surfaces are used in most serious computer graphics work. The *Maple* command

```
plot3d( [ f(u,v), g(u,v), h(u,v) ],
              u = a..b, v = c..d );
```

plots the parametric surface defined by

$$x = f(u, v), \ y = g(u, v), \ z = h(u, v)$$

for $a \leq u \leq b, \ c \leq v \leq d.$

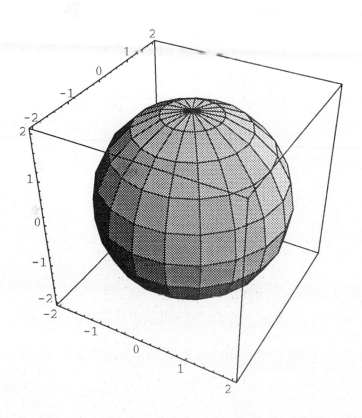

Example
The following code uses the spherical-coordinates equations

$$x = \rho \sin \phi \cos \theta$$
$$y = \rho \sin \phi \sin \theta$$
$$z = \rho \cos \phi$$

to plot a sphere of radius 2:

```
rho := 2;
plot3d( [ rho*sin(phi)*cos(theta),
           rho*sin(phi)*sin(theta),
              rho*cos(phi) ],
        phi = 0..Pi,   theta = 0..2*Pi,
        grid =   [11,20] );
```

The following projects list some possibilities for you to try for yourself.

1.   Use the cylindrical-coordinates parametrization

$$x = r \cos \theta, \quad y = r \sin \theta, \quad z = f(r,\theta)$$

of the surface $z = f(r,\theta)$ to plot some cones $(z = kr)$ and paraboloids $(z = kr^2)$. Then try something more exotic -- perhaps $z = (\sin r)/r$.

2.   Show that the elliptic cylinder $(x/a)^2 + (y/b)^2 = 1$ centered along the z-axis is parametrized by

$$x = a \cos \theta, \quad y = b \sin \theta, \quad z = z$$

for $0 \le \theta \le 2\pi$. Plot cylinders with varying semiaxes a and b.

3.   Show that the parametrization

$$x = ar \cos \theta, \quad y = br \sin \theta, \quad z = r^2$$

produces the elliptic paraboloid $z = (x/a)^2 + (y/b)^2$. Plot a few such paraboloids with different values of a and b.

4.   Show that the parametrization

$$x = a \sin \phi \cos \theta, \quad y = b \sin \phi \sin \theta, \quad z = c \cos \phi$$

produces an ellipsoid. Plot some with a few different semiaxes a, b, and c.

5.   Show that

$$x = x, \quad y = f(x) \cos \theta, \quad z = f(x) \sin \theta$$

for $a \leq x \leq b$, $0 \leq \theta \leq 2\pi$ parametrizes the surface obtained by revolving the curve $y = f(x)$, $a \leq x \leq b$ around the x-axis in xyz-space. Plot a few such surfaces generated by a variety of curves $y = f(x)$.

6. Explain how the standard parametrization

$$x = (b + a \cos \psi) \cos \theta, \qquad y = (b + a \cos \psi) \sin \theta, \qquad z = a \sin \psi$$

of a circular torus (obtained by revolving the center of a circle of radius $a < b$ around a circle of radius b) can be altered to graph a torus with an elliptical (rather than a circular) cross section, as pictured below. It might be more of a challenge to revolve the ellipse around an elliptical (rather than circular orbit).

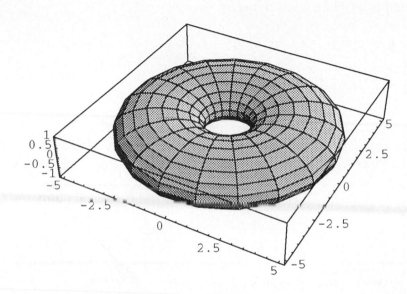

Chapter 16

Vector Analysis

Project 48
Computer Plotting of One-Sided Surfaces

Reference: Section 16.5 of Edwards & Penney

Suppose that we revolve a line segment around the z-axis in space while simultaneously rotating the segment in its own vertical plane. If the line segment rotates vertically through the angle $p\theta$ as it revolves through the angle θ around the z-axis, then the surface generated is a ribbon with p *twists*.

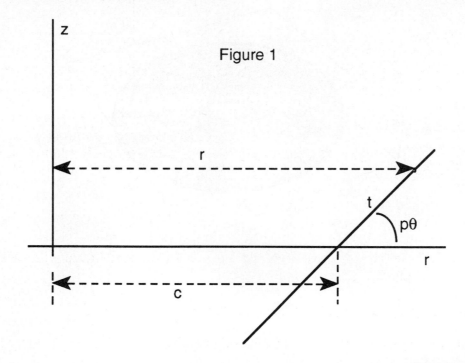

Figure 1

If the line segment has length 2 and its midpoint describes a circle of radius c as it revolves around the z-axis, use Fig. 1 to show that this ribbon is parametrized by the equations

$$x = (c + t \cos p\theta)\cos\theta,$$
$$y = (c + t \cos p\theta)\sin\theta, \tag{1}$$
$$z = t \sin p\theta$$

for $-1 \le t \le 1$, $0 \le \theta \le 2\pi$.

With $c = 4$ and $p = 1/2$ twists, you should get the one-sided Möbius strip on the left below. The figure on the right shows the ribbon constructed with a full twist $(p = 1)$. Can you see that it is two-sided? What happens if $p = 3/2$?

98

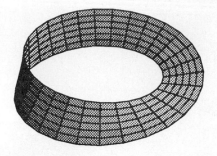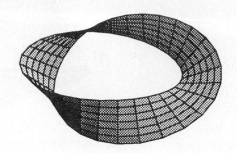

Twisted Tubes

When a closed curve is revolved and twisted, we get a *twisted tube* instead of a ribbon. Suppose the ellipse in the xy-plane with parametric equations

$$x = a \cos t, \quad y = b \sin t \tag{2}$$

through the angle α. Apply the formulas in Eq. (7) of Section 10.7 (rotation of axes) to show that the rotated ellipse is parametrized by the equations

$$x = a \cos t \cos \alpha - b \sin t \sin \alpha,$$
$$y = a \cos t \sin \alpha + b \sin t \cos \alpha. \tag{3}$$

Now suppose that an ellipse with semiaxes a and b is revolved with twist p around a circle of radius c. If

$$r(t, \theta) = c + a \cos t \cos p\theta - b \sin t \sin p\theta, \tag{4}$$

use the equations in (3) to show that the resulting twisted tube is parametrized by the equations

$$x = r(t, \theta) \cos \theta,$$
$$y = r(t, \theta) \sin \theta, \tag{5}$$
$$z = a \cos t \sin p\theta + b \sin t \cos p\theta$$

for $0 \le t \le 2\pi$, $0 \le \theta \le 2\pi$. When you parametrically plot these equations with $a = b = p = 1$ and $c > 1$, you should get an ordinary torus. The *cruller* shown in the next figure is obtained with $a = 3$, $b = 1$, $c = 7$, and $p = 3/2$.

99

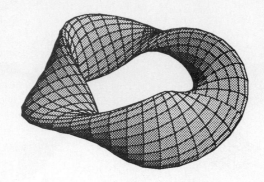

The Klein Bottle

The parametric equations

$$x = \sin t, \qquad y = \sin 2t \tag{6}$$

for $0 \le t \le 2\pi$ describe a figure-eight curve -- parametric plot it to see for yourself. If you use the equations in (6) rather than those in (2) to generate a twisted tube with twist $p = 1/2$, you should obtain the *Klein bottle* shown below. It is a representation in space of a closed surface that is one-sided. The true Klein bottle does not intersect itself; it is a surface much like a torus, but it is one-sided. It cannot exist in ordinary three-dimensional space. This is indicated by the fact that its representation in the figure intersects itself in a simple closed curve.

   If twisted tubes interest you, additional suggestions may be found in C. H. Edwards, "Twisted Tubes," *The Mathematica Journal* **3** (Winter 1993), 10-13.

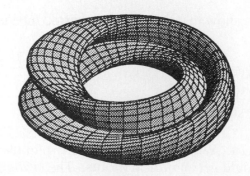